Make:

INVENTING A *Better* MOUSETRAP

*200 Years of American History in the
Amazing World of Patent Models*

Alan & Ann Rothschild

Introduction by Gareth Branwyn

Foreword by Forrest Morton Bird, M.D., Ph.D., Sc.D., D.S.

Inventing a Better Mousetrap
200 Years of American History in the Amazing World of Patent Models
By Alan and Ann Rothschild
Copyright © 2016 Alan and Ann Rothschild. All rights reserved.
Printed in Canada.
Published by Maker Media, Inc., 1160 Battery Street East, Suite 125, San Francisco, California 94111.
Maker Media books may be purchased for educational, business, or sales promotional use. Online editions are also available for most titles (http://safaribooksonline.com). For more information, contact our corporate/institutional sales department: 800-998-9938 or corporate@oreilly.com.

Editor: Gareth Branwyn, Emma Dvorak, and Brian Jepson
Photography: Scherzi Studios
Proofreader: Brian Jepson
Indexer: Brian Jepson
Interior Designer: Holly Scherzi Design
Cover Designer: Holly Scherzi Design

November 2015: First Edition
Revision History for the First Edition
2015-11-15: First Release

978-1-4571-8718-6

[TI]

THE UNITED STATES OF AMERICA

N.º 1092106

DEDICATION

This book is dedicated to the four hundred men and women inventors whose patent models are featured throughout this book. Some of the inventors became famous, such as B.F. Goodrich, Christian Steinway, Nelson Goodyear, and Linus Yale, Jr., known for patenting life-changing inventions. But for most of the inventors chronicled here, their hopes and dreams of becoming rich and famous, or creating some revolutionary new technology, never materialized. Their inventions remained in obscurity, until now. These inventors are finally being recognized for their ingenuity, their imagination, and their efforts in obtaining a United States Patent. May their memories live and be remembered into perpetuity.

We would also like to dedicate this book to our good friend, Dr. Forrest Bird (June 9, 1921-August 2, 2015). Dr. Bird's respiratory inventions have been responsible for saving hundreds of thousands of lives throughout the world. Not only was he a major inspiration to us, but his impact on our world will continue to be felt. Forrest Bird was a brilliant yet humble man who was one of the world's greatest inventors. We will miss him very much.

Alan & Ann Rothschild

Table of Contents

All of the patent models featured in this book are part of the Rothschild Patent Model Collection.

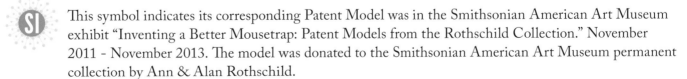 This symbol indicates its corresponding Patent Model was in the Smithsonian American Art Museum exhibit "Inventing a Better Mousetrap: Patent Models from the Rothschild Collection." November 2011 - November 2013. The model was donated to the Smithsonian American Art Museum permanent collection by Ann & Alan Rothschild.

 This symbol, "Makey" indicates its corresponding Patent Model is included in Chapter 25, Six Patent Models You Can Make.

The United States of America

TO ALL TO WHOM THESE LETTERS PATENT SHALL COME:

Ira C. Lucy, of Bernardston, Massachu

"The patent system added the fuel of interest to the fire of genius."

—*Abraham Lincoln*

Foreword

For ninety years the U.S. Patent Office required anyone who sought a patent to submit a model of their invention. Patent models represent American ingenuity at its finest. They represent the imagination that innovators have, and a fever for changing the world. Hundreds of thousands of models created a unique tangible record of human ingenuity unlike anything before or since.

Since many of the early inventors were common people without the technological, legal or commercial training that we have today, it was a difficult task for them to undertake to build a model not to exceed 12" by 12" by 12" to demonstrate the operation of their inventions. After the model requirement was abolished in 1880, America was in danger of losing its priceless record of innovation.

Fortunately for our America, a few historians who recognized the historic, scientific and artistic value of patent models had the vision to collect, protect and share our national treasure of patent models.

Alan and Ann Rothschild dedicated a significant part of their lives to preserving the history and legacies of their forefathers. The Rothschild Patent Model Museum is unmatched amongst private museums. They have given of themselves unto others and to history.

As the inventor of the Bird Respirator, I understood the challenges that inventors face on a daily basis and what they bring to the world. Inventors not only create jobs, they create industries.

My path and the Rothschilds' have crossed several times. I have been fascinated with their time, energy, efforts and ingenuity to refurbish, restore and nurture these precious patent models in history. The Bird Aviation Museum and Invention Center is particularly grateful to Alan and Ann for their gifts of historic examples of ingenuity to share with all people who walk through our museum and our lives. Their gifts have ever endeared the hearts of my wife, Pamela, and I as we watch people of all ages marvel at these rare pieces of our nation's history.

Alan and Ann, YOU have changed the world. You have preserved history for others to see. You have opened your arms, home and hearts and given of yourselves. I commend you. We forever thank you. God speed!

Forrest Morton Bird, M.D., Ph.D., Sc.D., D.S.
Inventor, Medical Respirator
National Inventor Hall of Fame Inductee

Introduction
A MODEL OF AMERICAN INGENUITY

Since I was a child, I've had an obsession with all types of models and the art of model-making. I built models throughout my childhood. I'm still building models today.

There is something undeniably compelling, almost magical, about seeing some element of the world: a machine, a vehicle, a building, a memorable scene from history captured in a diorama, all in life-like miniature detail. Peering closely at a well-produced, well-presented model is almost like peering into another world or a precisely frozen rendering of our own.

Patent models, those three-dimensional manifestations of a new idea, new mechanism, or an entirely new realm of technology, were once required to be submitted with all patent applications. These late 18th century/19th century models may have had a utilitarian purpose, but they are no less compelling or imbued with enchantment than any other type of model.

Patent models often say a lot about the people who made them (or could afford to have them professionally made). Among other things, they are an expression of the means available to the inventor. Some of them are quite crude, even whittled out of wood, while others pushed the limits of what could be machined and fabricated at the time. In many ways, these models encode the spirit of a young, optimistic America and its grand experiment in democratic governance, one that challenged people from all walks of life to invent their own future.

Our inventions say a lot about our desires, our dreams and aspirations for a better future. They express the best of us, our deep desire to improve humanity's lot. They also give an almost comical insight into our less noble side – our desires to "get rich quick" and to never get off the couch, relegating the messy business of sustaining our lives to the machinery we've invented to surround us.

Besides my love of modeling, in my career as a writer and editor focusing on do-it-yourself media and technology, I have always been keenly interested in garage-borne innovation and bursts of game-changing creative genius.

`These interests, model making and the nature of invention, perfectly converge in Ann and Alan Rothschild's *Inventing a Better Mousetrap*. Their impressive patent model collection (4,000 models and counting) forms the heart of this beautiful, informative, and entertaining book.

The Rothschilds not only share with us some of their more interesting and important models, and the models' often equally interesting histories, but they also offer insights into the spirit and specifics of the time, giving context to these inventions, some of which changed the course of American history and the literal engines that powered it.

Ironically, we have recently reentered the era of the design model. We now have 3D design programs and desktop 3D printers for rapidly visualizing and prototyping our ideas. A lot of the innovations around 3D designing, scanning, and printing have come to us from the so-called Maker Movement which has emerged over the past decade. This new identity as a "maker" – bestowed upon hobbyists, tinkerers, crafters, amateur robot builders, and anyone else who cares to assume the mantel—has inspired a new generation in search of innovative potential in America (as well as other regions of the world). In this "permission to play" ethos, with its open source sharing of ideas and project plans over the Internet, and through events like global Maker Faires, true and disruptive innovations are being incubated. The consumer desktop 3D printer, desktop CNC routing robots, and cheap and easy to understand and program tiny computers, like the Arduino and the Raspberry Pi, have all grown out of the Maker Movement.

In President Obama's 2009 inauguration speech, the ears of every maker listening pricked up when he issued a call to ". . . the risk-takers, the doers, the makers of things." It was these makers and risk-takers who forged a United States out of a liberated British colony. And who will now help forge the early 21st Century.

In the true hands-on spirit of making, *Inventing a Better Mousetrap* even includes six patent models that you can build yourself. The patent model was a way of making what might be a complex, hard-to-understand mechanism clearer to patent examiners and to the general public (many patent models were—and still are—on display at The Smithsonian, the U.S. Patent Office, and elsewhere). Seeing is believing. Using this book, you can get hands-on with these models of American ingenuity by building replicas of six submitted patent models.

And it is also in the spirit of DIY that Ann and Alan Rothschild are to be commended. They turned a personal passion for patent model collecting into an important historical project that benefits us all and comes to full fruition with this book.

A new generation of risk-takers, doers, and makers will hopefully be inspired by those who have come before and by the models and maps of their discovery process that they left for us—and that Alan and Ann have so thoughtfully preserved and present to you now.

Gareth Branwyn
Author and chronicler of DIY media and technology
Former Editorial Director of Make: magazine

We want to thank the following individuals and institutions
who have worked with us over the past twenty-one years.
You all have helped give our patent model collection its rightful place
in our country's history of progress and innovation:

The Albany Boyz

Drs. Forrest and Pamela Bird

Phil Bousquet

Gareth Branwyn

Betsy Broun

Dan Cichello

Cornell University

Jim & Diane Davie

Senator John DeFrancisco

Emma Dvorak

Rhea Fletcher

Karen Grossi

Gary Grossman

Brian Hoke

Brian Jepson

Maggie Johnson

Anne Jones

Kathy Kaminski

Don Kelly

Gary Kohs

Nell Lewis

Richard Maulsby

Massachusetts Institute of Technology

Kim Miner

Morrisville State College

Onondaga Community College

Cliff Petersen

Eileen Pezzi

Paul Predmore

Vicki Quigley

The Research Foundation
for the State University of New York

Charles Robertson

Dr. Michael Roizen

Neil Salkind

John Scheib

Mitch Scott

Holly Scherzi

Jim Scherzi

Paul Solomon

Paul "Bud" Soper

State University of New York at Cortland

State University of New York at
Oneonta Cooperstown Graduate Program

Peter Steinman

Syracuse University

Frank Teng

Rod Tyo

*And especially thank you to all our friends
and family who have shared and encouraged us
during this fascinating journey!*

Authors Alan and Ann Rothschild

The United States of America.

To all to whom these Letters Patent shall come :

WHEREAS *John Nazro* a citizen of the State of *Massachusetts* in the United States, hath alleged that he has invented a new and useful improvement *in extracting Mineral Alkali from marine Salt and Kelp, and making compound Pot and Pearl Ashes.*

which improvement has not been known or used before his application ; has *made oath* that he does verily believe that he is the true inventor or discoverer of the said improvement ; has paid into the Treasury of the United States, the sum of thirty dollars, delivered a receipt for the same, and presented a petition to the Secretary of State, signifying a desire of obtaining an exclusive property in the said improvement, and praying that a patent may be granted for that purpose : THESE ARE THEREFORE to grant, according to law, to the said *John Nazro* his heirs, administrators, or assigns, for the term of fourteen years, from the *thirty first* day of *october last past* the full and exclusive right and liberty of making, constructing, using, and vending to others to be used, the said improvement, a description whereof is given in the words of the said *John Nazro* himself, in the schedule hereto annexed, and is made part of these presents.

IN TESTIMONY WHEREOF, *I have caused these Letters to be made Patent, and the Seal of the United States to be hereunto affixed.*

GIVEN *under my hand, at the City of Philadelphia this sixth day of January in the Year of our Lord, one thousand seven hundred and ninety- Seven and of the Independence of the United States of America, the Twenty-first.*

G. Washington

By the President,

Timothy Pickering Secretary of State.

City of Philadelphia : TO WIT :

I DO HEREBY CERTIFY, That the foregoing Letters Patent, were delivered to me on the *fifth* day of *January* in the year of our Lord one thousand seven hundred and ninety- *seven* to be examined ; that I have examined the same, and find them conformable to law. And I do hereby return the same to the Secretary of State, within fifteen days from the date aforesaid, to wit : On this *sixth* day of *January* in the year aforesaid. *Charles Lee attorney genl*

*Figure 1: Original patent paper in the Rothschild Patent Model Collection.
Signed by George Washington, 1797.*

CHAPTER 1

History-Timeline

"The farther backward you can look, the farther forward you are likely to see."

— Winston S. Churchill

> **IT IS APRIL 10, 1790. PRESIDENT GEORGE WASHINGTON HAS JUST SIGNED A BILL TO CREATE THE UNITED STATES PATENT OFFICE. FOR THE FIRST TIME IN AMERICAN HISTORY, THE RIGHT OF AN INVENTOR TO PROFIT FROM HIS OR HER INVENTION IS RECOGNIZED BY LAW.**

- The Constitution (Article 1, Section 8, Clause 8) empowers Congress: "To promote the Progress of Science and useful arts, by securing for limited Times to Authors and Inventors the exclusive Right to their respective Writings and Discoveries."

- The subject matter of a United States patent is defined as "any useful art, manufacture, or device, or any improvement thereon not before known or used." To apply for a patent, a specification and drawing, and if possible, a model, are to be submitted.

- Secretary of State Thomas Jefferson, Secretary of War Henry Knox, and Attorney General Edmund Randolph, are chosen to head a three-member Patent Commission. Commission members are given the power to issue a patent—if they deem the invention or discovery sufficiently useful and important—for a period not to exceed fourteen years.

- The board's authority to grant patents is absolute, with no appeals process. The Department of State is given the responsibility for administering patent laws. The fee for a patent is between $4 and $5.

- July 1790, the first U.S. patent is granted to Samuel Hopkins of Pittsford, VT for a method of making potash and pearl ash by a new apparatus and process. Potash (potassium carbonate) is used in making soap and the manufacturing of glass.

- Every patent document issued between 1790 and 1836 is personally signed by the United States President, the Secretary of State, and the Attorney General. (Figure 1)

HISTORICAL TIMELINE

1790-1793

A total of 55 patents are granted to inventors. However, the board faces difficulties because its three members don't have the time to spare from their regular duties to sufficiently devote themselves to patent matters.

1793

The original patent law is revised. A simpler registration system is implemented, allowing anyone who applies and pays a $30 fee to be granted a patent. The patent board is eliminated, and the granting of patents falls to a clerk at the Department of State. This system remains in effect until July 4, 1836.

1794

On March 14, Eli Whitney (1765-1825) patents his invention of the cotton gin. Whitney's patent is the first significant patent to be issued by the recently reformed patent act. "King Cotton" becomes the major crop of the American South. Cotton soon represents more than half the total of all U.S. exports.

1800

On December 1, the United States Capital moves from Philadelphia, PA to Washington, D.C.

1802

James Madison, Secretary of State, creates a separate patent office within the State Department, appointing Dr. William Thornton as its first superintendent. Thornton's salary is $1,400 a year.

1809

The very first U.S. patent ever granted to a woman is issued to Mary Kies of Killingly, CT, on May 5th. Her patent is for a method of weaving straw with silk or thread.

1810

The Patent Office moves from the Department of State to Blodgett's Hotel, also known as the "Great Hotel" and "Union Pacific Hotel." It was built in 1793 by Samuel Blodgett on E Street between 7th and 8th Avenue. (Figure 2) Ironically the building is never used as a hotel. Instead, Washington's first formal playhouse, the "United States Theatre," opens in the building. For the first time, patent models are put on public display, and it becomes a local custom to stroll through the rooms on Sundays to see what new models are on view.

Figure 2: Blodgett's Hotel

1812

On June 18, the United States of America declares war on the United Kingdom of Great Britain and Ireland, its North American colonies, and its American Indian allies. The War is fought over maritime issues, the restriction of American trade with the European continent, and impressment, the Royal Navy's practice of removing seamen from American merchant vessels.

1814

The British burn government buildings in Washington, but the Patent Office is left untouched. Superintendent Thornton saves the Office by pleading with the British Commander not to "burn what would be useful to all mankind."

1815

On February 18, with the signing of the Treaty of Ghent, both parties return occupied land to its prewar owners and resume friendly trade relations.

1821

Thomas Jennings, born in 1791, becomes the first African-American to receive a U.S. patent. Jennings operates a laundry in New York City and patents an invention for a dry cleaning process.

1823

An attempt is first made to record and keep a list of all existing patent models. 1,819 are counted.

1836

The Patent Act of July 4 reestablishes the examination system of 1790. The submission of models are once again required by the Commissioner. *"The model, not more than 12 inches square, should be neatly made, the name of the inventor should be printed or engraved upon, or affixed to it, in a durable manner."* The requirement of submitting a model is a unique feature of the American patent system; no other country in the world requires or makes use of such models. The application fee is $30 for U.S. citizens, $500 for British subjects, and $300 for all others.

On July 13, a numbering system for issued patents is instituted, replacing the previous practice of using names. Patent No. 1 is issued to Senator John Ruggles of Maine for traction wheels on locomotive steam engines. Ruggles was the head of the committee to draft the new patent law.

On December 15, there is a fire in the Patent Office and the entire building burns to the ground. All the records and most of the models are destroyed. Congress appropriates $100,000 for the restoration of 3,000 of the most important ones. Luckily, enough records are held outside the office to allow almost all of the 10,000 patents to be reconstructed. They are given their original date and an X after their number. The X signifies that the patent was issued prior to July 1836.

Congress authorizes construction of a new building for the Patent Office. The site chosen is bounded by F and G Streets, between 7th and 9th Avenue. The design is a Greek Revival structure, modeled after the Parthenon in Athens. (Figure 3) Congress sets certain requirements for the new building: it must be fireproof, meet the requirements of the Patent Office for the next 50 years, and include galleries for displaying patent models to the public.

Figure 3: The Patent Office between F and G Streets

1840

The first wing of the new building is completed at a cost of $415,000. The North Wing is not completed until 1860. By the end of the 1840s, as many as 10,000 visitors per month came to view the models.

1842

"New and original designs" become patentable for a term of seven years. A "design patent" is a form of legal protection granted for the ornamental design of a functional item. Design patents have a separate numbering system. Design Patent No. 1 is granted to George Bruce of New York City for a collection of script and ornament typeface fonts.

1844

On June 15th, Charles Goodyear receives Patent No. 3,633 for the vulcanization of rubber. This very significant patent is responsible for many new products and eventually spawned entire industries.

1847

Thomas Edison is born on February 11. Edison, a prolific inventor, eventually receives 1,093 patents, along with thousands more from dozens of countries. Two of his most important inventions are the Electric Lamp (light bulb), patented in 1880, and the Phonograph or Speaking Machine, patented in 1878. Among his other well-known patents are the motion picture, the telegraph, and the telephone.

1849

Abraham Lincoln receives Patent No. 6,469 on May 22 for "A Device for Buoying Vessels over Shoals." Lincoln is the only U.S. president ever to receive a U.S. patent. His invention is never put into practical use.

1856

Nikola Tesla, undoubtedly one of the most famous and brilliant inventors in human history, is born on July 10. Tesla was a Serbian-American engineer and physicist who invented the first alternating current (AC) motor. He held 40 U.S. patents, mostly relating to the alternating current electrical system that he developed. He died broke in 1943 in New York City.

1861

On April 12, the U.S Civil War begins when the Confederate army Charleston Bay, SC opens fire on the federal garrison at Fort Sumter in Charleston Bay, South Carolina. The war between the Confederate States of America of the South, known as the "Confederacy" and the "Union" of the North is fought over the power of the national government to prohibit slavery in territories that were not yet states.

The term of a U.S. patent grant is extended from 14 to 17 years, and the 7 year extension is abolished. Foreign patent applicants now pay the same fees, and obtaining a patent becomes $35. These changes are implemented due to a dearth of patent applications resulting from the outbreak of the Civil War.

The Constitution of the Confederate States of America provides for the establishment of its own Patent Office.

1865

On March 6th, Abraham Lincoln's second inauguration as President takes place. Lincoln's inaugural ball is held in the Patent Office Building, on the third floor of the North gallery. Over 4,000 guests attend the gala event which includes dinner and dancing until four in the morning.

The war ends on May 9, 1865 with a Union victory and the abolishment of slavery. The Civil War is responsible for over 600,000 deaths and over 400,000 wounded on both sides.

By the end of the war, a total of 266 patents are issued by the Confederacy.

1870

Congress abolishes the legal requirement for models, but the Patent Office will keep the requirement anyway for another ten years.

1876

Because of space constraints, the public is finally barred from viewing patent models in the Patent Office.

1877

On September 24, a major fire breaks out on the first floor of the west wing of the Patent Office Building. The fire spreads quickly to the upper floors, destroying 12,000 rejected models. Another 114,000 models in the north and west halls are engulfed in flames. 87,000 models are totally destroyed and 27,000 are retrieved from the blaze. Congress appropriates $45,000 for their restoration. (Figure 4).

The model requirement is subsequently deemed impractical, so the law is changed to omit models unless required by the Commissioner. Of the 246,094 patents that had been issued by 1880, perhaps 200,000 are represented by models.

1893

The models are moved out of the Patent Office and placed in storage. By the turn of the century, some models are still being submitted with patents.

THE CONFLAGRATION.

Figure 4: Patent Office fire

1908

Congress decides to get rid of its patent models. The Smithsonian selects 1,060 models. An auction of some 3,000 more models that failed to receive patents sell for $62.18. The remaining 150,000 or so are placed in storage, finally ending up in an abandoned livery stable.

1911

On August 8, Patent No. 1,000,000 is issued to Francis Holton of Summit, Ohio, for a vehicle tire.

1925

It is estimated that from 1884 to 1925, $200,000 is spent in moving and storing patent models.

On February 13, no longer willing to continue to pay for storage, Congress appropriates $10,000 to do away with the stored models. The Smithsonian is given 2,500 additional models, some are returned to the original inventors or their relatives. On December 3, the remainder of the models are sold at a public auction in New York City to philanthropist Sir Henry Wellcome, the founder of the Burroughs-Wellcome Company (now part of Glaxo-Wellcome).

1936

Sir Henry Wellcome dies at the age of 82, his dreams of establishing a patent model museum having been dashed in the 1929 stock market crash.

The trustees of Wellcome's estate sell the models for $50,000 to Broadway producer Crosby Gaige. He in turn sells the collection to a group of businessmen for $75,000. This group forms American Patent Models, Inc.

1940

American Patent Models, Inc. declares bankruptcy.

1941

The models are acquired by O. Rundle Gilbert, an auctioneer, in a bankruptcy auction at Foley Square in New York City for $5,000. He moves the models to his home in Garrison-on-the-Hudson, NY. Over the years, Gilbert holds many auctions and thousands of the models are sold. A fire destroys an untold number of them in one of Gilbert's storage facilities. On a number of occasions, Gilbert attempts to sell the entire collection.

1973

Cliff Petersen, a designer and inventor within the aerospace industry, begins collecting patent models from Gilbert.

1977

On February 18, an Exhibition is opened at California State University Fullerton, CA. It is titled "American Patent Models 1836/1880." It features 70 models borrowed from several private collections.

1979

Petersen purchases Gilbert's remaining collection of patent models, some of which have not been uncrated since the original packing in 1926, for $500,000. Peterson prints an eight volume set of catalogs showing patent models for sale.

1981

January 17-May 16, the Mississippi State Historical Society features an exhibit of 23 patent models all by inventors from the State of Mississippi.

1984

A major exhibit, *"American Enterprise: Nineteenth-Century Patent Models,"* opens in January at the Cooper Hewitt Museum in New York City. Over 300 models are borrowed for the exhibit from the Smithsonian Museums, Cliff Petersen, and Deborah and Jason Friedman. A beautiful 138-page catalog is printed for the exhibit.

1990

Petersen donates 30,000 models and one million dollars to the United States Patent Model Foundation. He keeps approximately 5,000 models in his personal collection.

1991

The Cooper Union, New York, NY curates an exhibition of patent models from the collection of Cliff Petersen. Petersen had graduated at 19 years old from Cooper Union in 1943 with a degree in civil engineering.

THE ROTHSCHILD PATENT MODEL COLLECTION
A Brief History

1994

On a hot summer day in the middle of August, Alan Rothschild stumbles upon a group of patent models for sale at an Upstate New York antique show. Completely captivated by them, but not really knowing what they are, he purchases a few. The next day, now totally intrigued, he goes back and buys a few more. These become the beginnings of the Rothschild Patent Model Collection.

1995-2000

After much research, Rothschild starts adding to the collection by purchasing models from other collectors, auctions, and private sales. He eventually meets Cliff Petersen and purchases Petersen's personal collection. The entire contents of the Patent Model Museum in Fort Smith, AR is also purchased and added to the collection. The Rothschild Patent Model Collection, housed in a museum in Cazenovia, NY, becomes the largest private collection of viewable U.S. patent models in the world. It features more than 4,000 patent models, including the work of many women, foreign, and famous inventors.

2000

Starting in January, the exhibit *"Patent Pending"* opens at the Museum of Science and Technology in Syracuse, NY. The exhibit features 200 patent models from the Rothschild Collection. When the exhibit ends in June 2001, over 500,000 individuals have viewed the models.

2001

In July, the Rothschild Petersen Patent Model Museum (RPPMM) receives museum status in New York State and official designation as a 501(c)(3) corporation.

Beginning in May and continuing to the present, an exhibit of 53 models from the collection are on display at Disneyland Resort Paris in Paris, France.

In October, patent models from the collection form the basis for the U. S. Patent and Trademark Office's exhibit, *"School Days."*

The Collection is featured in articles in *The Boston Herald*, the *Journal of Antiques*, and the *National Post* of Toronto.

In November, The Collection is featured in a segment on *Home and Garden Television*.

2002

Patent Models of watch inventions from the collection are featured in the *International Wristwatch* magazine.
In February, the patent models are the centerpiece of an exhibit at the U.S. Patent & Trademark Office, entitled "Icons of Innovation." An article about the exhibit and the models from the collection appear in the *New York Times*.
In April, the Patent Model Collection is featured in a segment on the television program, *CBS News Sunday Morning* with Charles Osgood.

In June, the collection is included in a story on the History Channels program, *Modern Marvels*.

In August, the Rothschild Patent Model Collection hosts the 7th Annual Independent Inventors Conference, ICON 2002. Co-sponsored by the Rothschild Collection, the U.S. Patent & Trademark Office, the United Inventors Association, and Onondaga Community College, the conference attracts inventors and aspiring inventors from across the country and Canada. With a keynote address by the inventor of the medical respirator, Dr. Forrest Bird, the two-day conference is a resounding success. An exhibit from the Rothschild Collection of over 200 patent models of inventors from the Central New York area serves as a backdrop to the conference which is held at Onondaga Community College in Syracuse, NY.

2003-2004

An exhibit of sixty five patent models from the Collection is on display at the Federal Reserve Bank of Boston from October 2003-March 2004. The exhibit is titled "Icons of Innovation".

2005

Ann and Alan Rothschild visit Disneyland Resort Paris to view the exhibit of patent models from the Rothschild Collection on display in the Discovery Arcade.

2006

Models and stories from the collection are featured in articles in Forbes, *The Journal of Antiques and Collectibles*, and *Antiques and Auction News*.

2007

The Rothschild Patent Model Collection is the subject of articles in *The Christian Science Monitor, Robb Report, Chubb Collectors, Open Skies: The Inflight Magazine of Emirates, American Spirit: Daughters of the American Revolution, Steinway and Sons*, and *American Profile*.

2008

Patent models from the Rothschild Collection go on permanent display at the Bird Aviation Museum and Invention Center in Sandpoint, ID.

2009

A cover story article on Alan Rothschild, the museum and the collection is featured in the Albany College of Pharmacy and Health Sciences' Postscripts magazine. Alan Rothschild received a B.S. degree in pharmacy from the college in 1965.

2010

After spending several years of planning to create a National Patent Model Museum/Invention Center to house the collection, Rothschild finally abandons the dream of building a bricks and mortar building after realizing it is not feasible.

Once again, after much planning and investigating different possibilities for the collection, the final choice is to create a national traveling exhibit that will showcase the Rothschild Collection in museums throughout the country. Working with Smith Kramer Fine Art Services, a traveling exhibit entitled, "The Curious World of Patent Models," is organized. Contents of the exhibit include 58 patent models, documents, and educational materials. The first venue opens in February 2010 and travels to museums and universities throughout the country throughout 2014. This is the very first exhibit of patent models to ever travel throughout the country.

FEBRUARY 27 - MAY 9 The Curious World of Patent Models Exhibit, Louisiana Art and Science Museum, Baton Rouge, LA

MAY 1 - DECEMBER 31 Several models from the Rothschild Collection are on display at the See Science Center in Manchester, NH

JUNE 7-AUGUST 22 The Curious World of Patent Model Exhibit, Alyce de Roulet Williamson Gallery, Pasadena, CA

OCTOBER 31 - APRIL 10, 2011 The Curious World of Patent Models Exhibit, Museum of Texas Tech University, Lubbock, TX

2011

MAY 1 - JUNE 19 The Curious World of Patent Model Exhibit, Alden B. Dow Museum of Science and Art, Midland, MI

SEPTEMBER 24 - DECEMBER 31 The Curious World of Patent Models Exhibit, The Hudson River Museum, Yonkers, NY

NOVEMBER 1 Wired magazine article, "Timeless Machines," is published, about the SAAM exhibit.

NOVEMBER 11 The exhibit "Inventing a Better Mousetrap: Patent Models from the Rothschild Collection," opens at the Smithsonian American Art Museum (SAAM) in Washington, D.C. The exhibit is curated by Charles F. Robertson, past Deputy Director of the SAAM. The exhibit features 35 patent models plus various related documents including a copy of a patent signed by George Washington on January 6, 1790. The original document is in the Rothschild Collection.

NOVEMBER 26 The *Wall Street Journal* published *"Trying to Build a Better Mouse Trap,"* also on the SAAM exhibit.

DECEMBER 1 An opening reception for the exhibit is held at the SAAM. Prior to the reception, an hour long conversation is held with Alan Rothschild and Charles Robertson in the SAAM auditorium. Over 250 guests and friends attend the gala event.

The title of the exhibition comes from the adage "Build a better mousetrap and the world will beat a path to your door," which is attributed to Ralph Waldo Emerson. And yes, there is a mousetrap in the exhibit. The inventors were Kopas & Bauer from Washington, D.C. Their invention, from 1870, was for a live trap that could actually catch four mice before being reset.

The SAAM was originally the building that was built for the Patent Office with construction beginning in 1836.

The exhibit remains open until November 7, 2013.

DECEMBER 8 The *Washington Post Express'* article on SAAM, "Patented Brilliance," is published.

DECEMBER 9 The *New York Times* runs the SAAM piece, "Think Big, Build Small: Inventors' Prototypes."

2012

JANUARY 22 - APRIL 1
The Curious World of Patent Models Exhibit, The Museum of Mobile, Mobile, AL.

FEBRUARY
American History publishes the story "The Very Model of a Better Mousetrap," on the SAAM exhibit.

APRIL 21 - JULY 1
The Curious World of Patent Models Exhibit, Clay Center for the Arts & Sciences, Charleston, WV

JULY 22 - SEPTEMBER 30
The Curious World of Patent Models Exhibit, Museum of Science and History of Jacksonville, Jacksonville, FL

AUGUST 18
Glancing Askance, *"Inventing a Better Mouse Trap,"* article on SAAM exhibit.

SEPTEMBER
Seoul, Korea's Chunjae Education magazine runs a story on the SAAM exhibit.

SEPTEMBER 29 - 30
Maker Faire New York 2012 exhibit at the New York Hall of Science, Queens, NY. "Building a Better Mouse Trap: American History in the Amazing World of Patent Models." The exhibit of 48 patent models is awarded a "Maker Faire Editor's Choice Blue Ribbon."

OCTOBER 21 - DECEMBER 30
The Curious World of Patent Models Exhibit, R.W. Norton Art Gallery, Shreveport, LA

2013

JANUARY 20 - MARCH 31
The Curious World of Patent Models Exhibit, Fullerton Museum Center, Fullerton, CA

APRIL 21 - SEPTEMBER 29
The Curious World of Patent Models Exhibit, Historic Arkansas Museum, Little Rock, AR

OCTOBER 18 -DECEMBER 28
The Curious World of Patent Models Exhibit, City of Lake Charles, Lake Charles, LA

2014

JANUARY 17 - MARCH 30
The Curious World of Patent Models Exhibit, Texas A & M University, College Station, TX

MAY 4 - AUGUST 24
The Curious World of Patent Models Exhibit, Grout Museum of History and Science, Waterloo, IA

2015
See Afterword

Figure 1: The Head-Shaped Boiler on Giovanni Branca's Steam Turbine

CHAPTER 2

Steam

"Those who admire modern civilization usually identify it with the steam engine and the electric telegraph."

— George Bernard Shaw

Mobility is one of the great engines of civilization. For much of human history, that mobility and engine power were provided by beasts of burden, the wind, and human muscle. With the effective harnessing and actuating of steam, human endeavor was able to take a giant leap.

The roots of steam power actually stretch back to the first century A.D., to the Greek mathematician, engineer, and inventor, Hero of Alexandria. Hero invented the aeolipile, or what has come to be known as the Hero engine. This contraption centered on a hollow metal ball, with 2 teakettle-like spouts protruding from it. The ball was hung between 2 pivots, allowing it to freely revolve. When steam from a boiling pot beneath it was forced into the ball through the hollow pivots, it jetted out through the spouts. The force of these 2 jets of steam, on opposite sides of the ball, made the engine spin.

Seventeen hundred years later, in Italy, Giovanni Branca's book, Le Machine (1629) featured, amongst its wood-engraved illustrations of 63 inventions, a turbine wheel powered by a jet of steam. With the engine he'd designed, he planned to use it to pound drugs into powder. Steam for the engine was made in a boiler shaped like a man's head and chest. It spewed forth from the man's mouth, struck a small paddle wheel, and the wheel was spun around. (Figure 1)

The axle of the wheel on Branca's contraption included 2 weights that rose and fell. Each time they fell, they struck the drugs housed in a pair of bowls. Basically, what he envisioned was a giant, steam-powered mortar and pestle set. As far as we know, the device was never built.

By 1650, demand for a new kind of power was critical. Coal mines in England were filling up with ground water which needed to constantly be pumped out. Previously, this job fell upon draft animals which were used to extract large buckets of water from the mines. The process was labor-intensive and expensive. Inventors began to wonder how they might harness steam instead of horses.

Though debated by some, England's Edward Somerset is commonly thought to have invented the first, crude prototype of what we think of as a steam engine, at least on paper. He detailed the device, calling it a "water-commanding engine," in his 1655 book of inventions. The steam-powered device could force water through a pipe, for a distance of up to 40 feet, i.e. creating a steam-powered sump or irrigation pump.

But experimenting with steam was dangerous. Boilers blew up, pipes burst under intense pressure, and injuries were frequent and often serious.

French physicist Denis Papin, while conducting a series of steam-driven piston and cylinder experiments, ended up inventing a safety valve, in 1680, which instantly made steam boilers more trustworthy and less dangerous. Papin's valve allowed steam to escape whenever there was more of it than the boiler could safely handle.

In 1690, Papin (who was an assistant to Christiaan Huygens, the great Dutch mathematician and astronomer) invented a proper piston and cylinder-driven steam engine. Papin's cylinder was a metal tube, his piston a metal disk that fit snugly inside. When steam was blown into the bottom of the cylinder, the piston was forced upwards. When the steam cooled and changed back into water, the piston was drawn back down. This up and down motion of his "steam digester" proved extremely useful in pumping flooded mines.

In 1698, Thomas Savery patented the first working steam engine, calling it a "Fire Engine." It was the first practical, commercial steam pump or steam engine. It was thought of as a piston-less pump because it had no moving parts. As soon as it was built, it was pressed into service pumping water out of the deep mines. It worked by a two-valve system, one supplying steam and one water. The steam valve was first opened to admit steam into a chamber, then the steam was condensed by admitting water from the other valve, creating a partial vacuum which effectively pulled the water up, creating an effective pump. While Savery's pump did work, it was unreliable and subject to explosions and steam leaks.

Using a piston and cylinder design, Thomas Newcomen, an English blacksmith (1663-1729), further refined Savery's designs, applying the concepts of condensation of steam to create a vacuum and atmospheric pressure to push the piston into the cylinders.

In 1712, Newcomen fastened a steam-driven piston to one end of a huge seesaw. The other end was connected by a long rod to a pump deep inside of a coal mine. Each time the piston came down, the rod moved up, bringing with it large quantities of water. Newcomen's engine saved the mines. The device had its drawbacks. It was slow, it took a long time for the steam to cool, it required a lot of refueling, and it was inefficient. But Newcomen's engine was still an important beginning.

James Watt of Edinburgh, Scotland is the man most associated with the birth of the steam age. He produced his first commercial engines in 1776. He moved the steam engine to new levels of performance, using the pressure of steam to move pistons, first in only one direction, and eventually, up and down. Watt also incorporated a separate condenser into which used steam was diverted for condensation. This provided greater efficiency and cost-saving as the cylinder no longer had to be subjected to great changes in temperature with resultant heat loss. Watt's piston–driven engine attempted to improve on Newcomen's, inventing an engine that was much more powerful and faster.

Watt also devised a system to help mine owners decide how big of an engine they might need. He knew how much work a single horse could do, so he called that amount "one horsepower." He would then ask how many horses would be needed to do a job. If the mine owner thought he had 50 horses worth of work, Watt would make him a 50 horsepower engine. It is through Watt we get the term horsepower.

When Watt connected his piston to a wheel, it became evident that the steam-driven wheel could replace wheels traditionally powered by water. This was a major change. Factories and mills no longer had to be near a water supply. Power could be generated wherever it was needed.

Oliver Evans is sometimes known as the James Watt of America. Evans believed that a high-pressure engine could be half the size and weight of a low-pressure one and deliver more power using half the fuel. He built the first steam engineering factory for the production of steam engines. In 1811-12, with his son, George as partner, he opened the Mars Works, the first steam engine manufactory in Pittsburgh, PA. They built 100 stationary high-pressure steam engines for American workshops throughout the Atlantic states.

Evans made it possible for a steam engine to transport a profitable cargo in addition to its own weight. The Evans engine was known for its lightness and convenience. Evans was the first of a line of American inventors who helped bring the steam engine to prominence in the United States and he was also the first American to apply it to industry.

In the early years of the 19th century, the high-pressure steam engine was widely used, first on steamboats, and later in factories. Around the middle of the 19th century, the engine's basic components underwent significant development. For example, the shape and especially the construction of the piston were modified to meet the introduction of high pressures. By 1876 the "Age of Steam" was upon us.

Steam power soon dominated the railroad. The sailing ship was yielding to the steamer in passenger service, and to some extent, in freight. The American Navy had committed itself to steam power. And the development, by midcentury, of efficient and dependable stationary steam engines led to the explosive growth of industry. Single-acting engines, designed to achieve the lowest cost of operation and maintenance, continued to be used for drainage until the end of the 19th century.

Thousands of patents were issued to improve the efficiency of the steam engine. The history of this technology, as with many inventions, advanced through constant improvement and simplification of previous designs. The technological advances made to the steam engine would prove essential to powering America's great industrial revolution.

GEORGE CORLISS (1817-1888)

An American, George Corliss, is linked to James Watt in the historical development of steam power, even though his invention came 100 years later. The basis for Corliss' fame was his invention, in 1848, of the "automatic drop cut-off," an improved method for controlling the amount of steam admitted into a cylinder.

To Americans, the name Corliss was virtually synonymous with stationary steam power. (Figure 2).

Born the son of Dr. Hiram Corliss in Easton, NY, George grew to find world wide fame for his contributions to the design and construction of the steam engine. In 1847, he established Corliss, Nightingale & Co. in Providence, RI. By this time, he had already perfected his famous automatic drop cut-off invention. The first 2 engines featuring his device were completed in 1848 and proved highly successful. He was granted a patent in 1849. Corliss went on to devise many improvements to his engines, all of which were patented. In 1856, he

Figure 2: Portrait of George Corliss

moved to a new factory and continued production at a greatly expanded scale under the name of The Corliss Engine Co. Corliss received numerous awards for his inventions.

In 1867, Corliss was given the Paris International Exposition Award, and in 1868, he received the Montyon Prize from the Institute of France. When Corliss was awarded the Rumford Medal, in 1870, it was stated, "No invention since Watt's time has so enhanced the efficiency of the steam engine."

During his lifetime, Corliss received 70 patents, covering steam and pumping engines, machine tools, boilers, and a variety of technical appliances. Ahead of his time, he believed that in order to manufacture economically, the engines should be made by single-purpose machinery.

In 1880, he designed and built such special machinery in his shop.

He standardized engine sizes, arranged special machinery for sequential operations, and formulated plans for mass production and the interchangeability of parts.

Finding it difficult to convince businessmen that his engine was superior, Corliss issued a challenge to prospective buyers. He offered to give his engines away for free in exchange for the fuel savings his customers gained from his patented improvements. He sold one of his first engines with the understanding that he was to be paid all of the money it saved in 5 years. At the end of those years, he had pocketed $19,732, several times the price of the engine.

Corliss profited not only as a manufacturer, but also by selling licenses to other builders. Inevitably, he had imitators. There were also inventors who claimed that Corliss had pirated their designs. As a result, he was subject to almost continuous patent litigation and spent lots of time and money in defense of his inventions, his legal fees running well over one hundred thousand dollars. One of his lawyers was William H. Seward, who later became Secretary of State under Abraham Lincoln. Corliss' claims were sustained by the courts. After the expiration of his basic patents in 1870, dozens of builders embraced his principles, not only in the U.S., but around the world. By the end of the century, the "Corliss Engine" was the standard.

Of all the thousands of steam engines built at the Corliss factory, the most famous was the one at the 1876 Centennial Exhibition in Philadelphia. The Corliss steam engine was the primary symbol of the Centennial. The Exposition was sponsored by an act of Congress to celebrate the 100th anniversary of American Independence. Mr. Corliss was made a member of the Centennial Committee by President Grant and was also a member of the Executive Committee. Early in the planning, he offered to build a steam engine to power all of the machinery in Machinery Hall. Skeptics doubted this could be done. In less than 10 months, he designed and constructed the largest steam engine in the world.

The steam engine supplied power to 14 acres of machinery comprising several thousand machines in the building. The engine towered more than 40 feet in the air and weighed over 600 tons. The connecting rods were 24 feet long. The driving pinion, 10 feet in diameter, meshed with a fly wheel which was 30 feet in diameter. Six-foot bevel gears at spaced intervals drove cross-shafts, and in this manner, power was delivered to the entire building. The pinion shaft was 352 feet long.

President Grant and Emperor Dom Pedro II of Brazil opened the Exposition by operating the cranks and levers to start the engine located directly in the center of Machinery Hall. The engine ran continuously for 6 months without faltering.

Corliss also exhibited the gear cutter he designed to cut the teeth of the bevel gears that transmitted the Centennial Engine's power throughout Machinery Hall. (Figure 3)

In 1880, George Pullman decided the Corliss Engine from the Centennial Exhibition was just the thing to power the machinery in his new car works, then under construction near Chicago. It took a train of 35 cars to haul it to Pullman, IL. Pullman had an impressively appointed engine house erected for it and there it worked for nearly 30 years until 1910.

The Corliss engine became the standard by which other engines were measured.

Figure 3: Corliss' Impressive Bevel Gear Cutting Machine 1876

James Watt, Thomas Newcomen, and George Corliss are viewed as the great triumvirate in steam engine design and development. It was Corliss' role to revolutionize steam engine practice throughout the world.

IMPROVEMENT IN GEAR-CUTTING MACHINES, PATENT NO. 190,470
George H. Corliss, Providence, Rhode Island, May 8, 1877

George H. Corliss' impressive gear cutting-machine patent no. 190,470 was an improvement over an earlier patent no. 6161. This improved version was what he called "intrinsically superior" to the original. This machine was adapted to cut large bevel gears with any required degree of bevel. The cutting action of the machine was done with a "planing" operation as opposed to the common "milling" of the metal. The planing tool or "cutter" was adjustable at various positions to achieve the "inclination" or pitch required via a large moveable "trunk". This "trunk" was positioned up and down while being guided by a large curved cast iron 90 degree arch by use of ropes and pulleys. A rather large "index wheel", which the flooring had to allow it to extend below, was revolved by a hand wheel. This "index wheel" was used to fine tune the position, or turning of the currently worked gear to the present tooth being formed. This is the machine that was on display at the 1876 Centennial Exhibition.

JOSEPH HARRISON JR. (1810-1874)

Joseph Harrison, Jr. owned and managed Harrison Boiler Works of Philadelphia, PA. The company was employed exclusively in making a steam boiler of Harrison's invention using new principles of construction. A significant claim was that the boiler was absolutely safe from explosion. As Harrison says in his 1867 *An Essay on the Steam-Boilers*:

"It is formed of a combination of cast-iron hollow spheres, each eight inches in external diameter, connected by curved necks, with rebate/rabbet machined joints, held together by wrought iron bolts, with caps at the ends. Every boiler is tested by hydraulic pressure at 300 pounds to the square inch. It cannot be burst under any practicable steam pressure. Under pressure which might cause rupture in ordinary boilers, every joint in this becomes a safety valve. No other steam generator possesses this property of relief under extreme pressure, without injury to itself, and thus preventing disaster. It is not seriously affected by corrosion, which soon destroys the wrought-iron boiler. Most explosions occur from this cause. It has economy in fuel, equal to the best boilers. Any kind of fuel may be used under this boiler, from the most expensive to refuse coal dust. It produces superheated steam without separate apparatus, and is not liable to priming or foaming. It is easily transported, and may be taken apart so that no piece need weigh more than eighty pounds."

Billed as a "Safe Steam Boiler," Harrison said his *"new Steam Generator combines essential advantages in SECURITY FROM EXPLOSION; in first cost; economy of fuel; and transportation."* (Figure 4)

The Harrison Boiler received the Rumford Medal, awarded by the American Academy of Arts and Sciences in Boston, on January 9, 1872. It was also awarded the First Class medal at the Worlds Fair, London, 1862 "for originality of design, and general merit." It received the first medal and diploma at the American Institute Fair, New York, 1869. And at the American Institute Fair, two Harrison Boilers were exhibited and were the only boilers on the grounds that were found reliable and capable at all times during operation.

The judges, in the Centennial Exhibition, 1876 report, noted that:

"In the Harrison boiler great pressures of steam can be carried with great safety, that the great excess of strength in proportion to strains renders explosion in a high degree improbable; and that the small discharge of water set free, when any parts in combination give away, renders disastrous explosions also in a high degree improbable."

STEAM GENERATORS, PATENT NO. 80,543

Joseph Harrison, Jr., Philadelphia, Pennsylvania, August 4, 1868

Joseph Harrison Jr. patented an improvement to steam boilers, # 25,640 on Oct. 4, 1859. This original patent was a novel way to simplify, facilitate, strengthen, and cheapen the construction of boilers, resolving them into what he would call "units of construction." This method of boiler construction enabled one to easily replace sections that were damaged due to heat or other physical impairment. These units could be pieced together in any given shape or dimension. Its distinct spherical globe and tube construction was designed with strength under pressure in mind. In 1868, Harrison patented a new improvement to his previous 1859 patent. Patent # 80,543 improved on his older design by implementing the use of "compensating units". Because there is always one portion of the boiler unit that takes on greater heat, this would create an unequal expansion in the units. This unequal expansion can subject some of the closely connected units to undue strain and in some instances lead to fracture. To obviate this, the placement of an "elongated unit" integrated per his design, would relieve this pressure by independently expanding from the others.

TRUCKSON LaFRANCE (1834-1895)

The American LaFrance Fire Engine Company was one of the oldest fire apparatus manufacturers in America, building hand-drawn, horse-drawn, and steam-powered fire engines. The company was founded in 1832 by Truckson LaFrance and his partners as the LaFrance Manufacturing Company which produced hand pumps and rotary steam engines based on LaFrance's patents. As his designs began winning major national competitions, the LaFrance name began to spread. The Elmira plant was formed in 1872 by brothers, Asa W. & Truckson S. LaFrance. Truckson was the mechanical genius, Asa the master salesman. The company's history shows the manufacture of many things, including rail-fence machines, cotton pickers, corn shellers, and steam locomotives.

In late 1873, the LaFrance Manufacturing Company built the first rotary steam fire engine, which was soon sold to the city of Elmira. The engine had one weakness: the cams on the pumps wore down, finally refusing to deliver the needed pressure. Firemen fixed this fault by pouring molasses over the cams. For years, a jug of molasses was regular equipment on the old "LaFrance." In 1882, the company signed an agreement with Daniel Hayes to begin production of his screw-driven extension ladder truck which extended 85 feet above the ground. In 1884, LaFrance introduced a new engine design, a piston steam fire engine. (Figure 5)

In January 2014, after 180 years in business the company closed its doors.

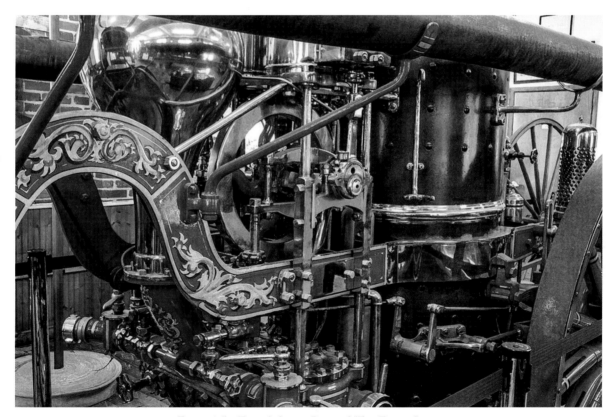

Figure 5: La France's Steam-Powered "Fire Engine," 1911

STEAM GENERATORS, PATENT NO. 108,604
Truckson S. LaFrance, Elmira, New York, October 25, 1870

La France's improved Steam Generator was designed to utilize as much heat energy from the device's firebox. The inclusion of "superheating spaces" and "smoke chambers" were designed to extract as much steam-producing heat energy as possible before its inevitable expulsion. The inclusion of water pipes directed into the firebox and unique water and steam space arrangement were meant to insure good water circulation as well.

FIRE ENGINE BOILER, PATENT NO. 231,336
Truckson S. La France, Elmira, New York, August 17, 1880

This fire-engine boiler design was to better protect the "crown sheet." The crown sheet is the highest heating surface in the boiler, so damage or explosion could occur if the water line dropped below it. In this new design, the extension of the walls of the fire-box around and above the crown sheet caused the retention of a large quantity of water upon the crown sheet, thereby protecting it. This safety feature would give the boiler operator the much needed time to re-supply the vessel with water. Another advantage of this design was the ability to tap the superheated water from the crown sheet vessel, allowing the operator to get up a quick head of steam when needed. The addition of a new "mud drum" added a well which could catch a large percentage of sediment and scale delivered by the incoming water supply.

 GOVERNOR FOR STEAM AND OTHER ENGINERY, PATENT NO. 99,751
Joseph Bell, Cincinnati, Ohio, February 12, 1870

An invention that improves the control of the mechanism that regulates the revolutions per minute made by the engine's crank-shaft.

GOVERNOR FOR STEAM AND OTHER ENGINES, PATENT NO. 25,769
H.D. Snow, Rochester, New York, October 11, 1859

The adjustable stop or collar controls the passage of steam through the valve on starting the engine, and also controls the extreme downward movement of the valve.

SURFACE BLOW-OFFS, PATENT NO. 154,734
Robert Waugh, New Orleans, Louisiana, September 1, 1874

A hollow flat skimmer that thoroughly removes the scum which rises to the surface of the boiling water. A steam boiler can run four or five times longer than a boiler without this mechanism before needing to be cleaned out.

INSTANTANEOUS GOVERNOR FOR STEAM-ENGINES, PATENT NO. 14,967
Wm. W. H. Mead, Chestertown, New York, May 27, 1856

A novel combination of a centrifugal governor or fly with a throttle valve. The movement of the inertia of the governor acts upon the valve in such a way that a tendency toward increased engine speed diminishes the supply of steam, and vice versa.

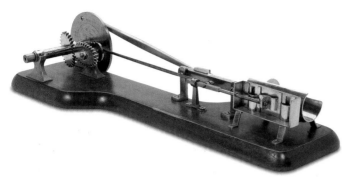

GEARINGS FOR VALVE-MOVEMENTS FOR STEAM-ENGINES, PATENT NO. 132,130
William S. Bacon, Sulphur Springs, Ohio, October 15, 1872

A set of improvements in valve-movement for steam engines, using elliptical-toothed gears. By constructing and arranging the valve-gear to produce a movement, steam can be used expansively with a single valve.

COMBINED GAS, AIR, AND STEAM ROTARY ENGINE, PATENT NO. 249,214
William H. Wigmore, Philadelphia, Pennsylvania, November 8, 1881

The engine introduces gas, air, and water simultaneously and together into a tank connected with a cylinder of a steam engine. The water turns to steam, which then combines with the heated air and gas to drive the engine.

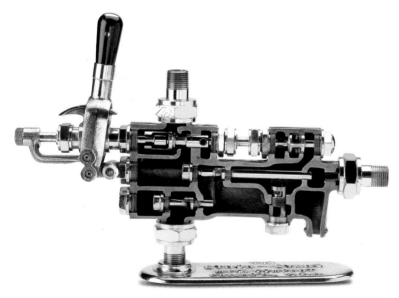

INJECTOR, PATENT NO. 587,279
Thomas M. Eynon, Joseph W. Gamble, Philadelphia, Pennsylvania, April 9, 1895
Assignors to The Eynon Evans Manufacturing Company, of Pennsylvania

For this improvement to the double tube injector, "our object is to construct a compact and symmetrical injector, embodying all the essential features of Patent No. 497,269, granted to Thomas E. Eynon and to incorporate additional features."

Please refer to the Patent Papers (http://www.google.com/patents/US537279) *to read the 18 additional features that are claimed in this complex steam injector invention.*

TEMPERATURE-REGULATOR, PATENT NO. 457,280
Adam Kelly, Smithfield, Rhode Island, August 4, 1891

An apparatus that regulates the temperature of spaces heated by steam, hot water, hot air or other medium delivered through pipes and conduits. The ratchet-wheel controls a valve-operated lever and a pivotally-mounted mercury thermometer tube.

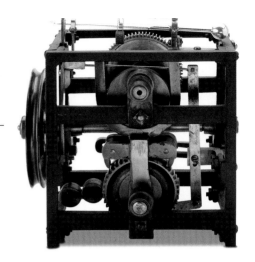

ROTARY ENGINES, PATENT NO. 184,919
Josiah M. Simpson, Oshkosh, Wisconsin, November 28, 1876

Simpson's "Improved Revoluble Reciprocating Steam-Engine" is an example of a particularly elegant rotary engine. In this invention, the rotating cylinders act as their own sliding valve to distribute steam on the 2 sides of the piston with moving parts kept to a minimum.

STEAM-JET PUMPS, PATENT NO. 206,054
Hanson P. Tenant, East Germantown, Indiana, July 16, 1878

An invention that facilitates the construction and installation of steam-jet pumps. The steam vents are made separately from the pump, with a shell or globe cast around them for a steam-tight joint. This pump is entirely free of bolts and is "so simple that an ordinary gas-fitter can set it up in a very short time and without the possibility of misapplying any of its parts."

AUTOMATIC STOP-MOTIONS FOR STEAM-ENGINES, PATENT NO. 47,359
John Jackman Jr., Newburyport, Massachusetts, April 18, 1865
Assignor to The American Automatic Stop Motion Company

A spring and suitable stop that lets the engine run normally, unless the governor-ball drops, which liberates the spring and automatically stops the engine.

STEAM-TRAPS, PATENT NO. 143,761
James W. Hodges, New York, New York, February 21, 1873

This invention describes a tubular slide-valve operated by a float, and is balanced such that it operates with equal freedom under any pressure. The slide-valve and float are applied in a metallic case that has screw-plugs or valves for regulating the speed of the discharge water of condensation, and for blowing out obstructions.

APPARATUS FOR REGULATING THE SUPPLY OF WATER TO STEAM-BOILERS, PATENT NO. 21,003
Zalmon L. Jacobs, Hebron, Connecticut, July 27, 1858

The nature of this invention consists of a combination of a chamber having alternate communication, with a reservoir to receive a fluid, and a boiler or other vessel in which to deliver it. This causes the fluid, when it rises to the desired height in the vessel, to check the passage of air to the chamber, thereby automatically regulating the flow of fluid.

ROTARY STEAM-ENGINES, PATENT NO. 2,015
Jesse Tuttle, Boston, Massachusetts, March 26,1841

Tuttle invented a new and useful Improvement in Rotatory Steam Engines and called it "Jesse Tuttle's Patent Rotatory Engine." The operation of the engine worked by admitting steam into the steam-chest, which moves the side-valve to open, which communicates with the piston to rotate. The steam then rushes through its port into the open side pipe then into the steam- chamber. The steam then travels through the steamway in the rotary plate into the annular chamber which ends up forcing a continuous rotary motion.

REVOLVING-CYLINDER ENGINES, PATENT NO. 51,166
Joseph S. Foster, Virginia, Nevada Territory, November 28, 1865

This invention relates to an improvement in that class of steam-engines in which one or more cylinders are attached to a revolving disk. In this new engine two cylinders are rigidly attached to a revolving disk, placed opposite each other in a radial direction. There is a common piston rod attached at the center to a crank-pin, in such a manner that every revolution of the fly-wheel (caused by the action of the steam in the cylinders) produces two revolutions of the crank which is transmitted to a working machine.

MAGNETIC STEAM-GAGE, PATENT NO. 20,851
Joshua Lowe, New York, New York, July 6, 1858,

Lowe's invention consists of combining a tight chambered mercury gauge for steam-pressure (or vacuum) with self-adjusting (or floating) magnets, magnetic needles, and index dials. These are combined in such a manner as to produce magnetic mercury pressure and vacuum gauges that will give true magnetic indications of varying degrees of pressure occurring within steam boilers, or of vacuum in condensers.

ROTARY STEAM ENGINES, PATENT NO. 141,436
Dexter D. Hardy, Delavan, Illinois, August 5, 1873

Dexter D. Hardy patented improvement to rotary steam engines consisted in arranging and constructing the devices' workings to obtain a "uniformity of wear" throughout the machine. Their eccentrically-placed center shaft and interior drum design were but two of the highlighted improvements in this style of engine.

CHAPTER 3

Alcohol, Tobacco & Firearms (ATF)

"Sir, if you were my husband, I would poison your drink."
"Madam, if you were my wife, I would drink it."

— Exchange between Lady Astor and Winston Churchill

ALCOHOL

Alcohol has played an important and frequently disreputable role throughout the annals of human history. This was certainly the case in colonial America. The Puritans are widely reported to have brought more beer aboard the Mayflower than water. This seems counter to our perceptions of puritanical behavior until you realize that drinking wine and beer was often safer than drinking water. (The much-repeated story goes, whether true or not, that the Mayflower rushed to find its landing place after the beer supply ran out.) Beer is likely to have been the first alcoholic beverage produced at Jamestown and Plymouth.

So, it is perhaps also no surprise that whisky-making soon became a cottage industry in the colonies. The first liquors made use of a wide variety of ingredients: berries, plums, apples, carrots, grain, and potatoes. Until the mid-18th century, whiskey was made in relatively small batches, mainly by farmers-turned-distillers. Even George Washington was a distiller, producing his own rum and whiskey at Mt. Vernon in the 1770s.

Spirits have also long been the subject of government taxation. In 1791, Alexander Hamilton, then Secretary of the Treasury, proposed a tax to help the new country pay off its debts after the Revolutionary War. George Washington approved an excise tax on all spirits—both imported and domestic. The alcoholic strength of the product was used to determine the tax rate, and spirits made from natively-grown ingredients were taxed less than those made from imported goods. An annual tax was also levied against each still, based on its capacity.

Not surprisingly, the 1791 whiskey tax was unpopular amongst the farmer-distillers. In Pennsylvania, whiskey makers even revolted. In 1792, the government eased taxes somewhat, from eleven cents a gallon to around seven. After further reducing taxes, and still not getting cooperation from the Pennsylvania farmers, in 1794, Washington called in federal troops to quell the uprising. This intervention became known as the Whiskey Rebellion.

In the aftermath, pardons were granted to anyone who agreed to comply with the law going forward. Washington even offered incentives for pioneers who were willing to move southwest to the frontiers of Virginia, beyond the Allegheny mountains. Famously, Thomas Jefferson offered his own incentive: 60 acres to "create a permanent structure and crops of native origin." This led to the formation of Bourbon County, KY (known today as Clark County, KY), then a rich agricultural area on the frontier edge of American westward expansion. Pennsylvania farmers, many of Scottish and Irish descent, headed to Bourbon County to take advantage of the offered incentives.

The load on distillers was lessened somewhat when, in 1802, the Excise Act was repealed by President Thomas Jefferson. Apart from the few years between 1813 and 1817, when excise taxes were again levied in order to pay for the costs of the War of 1812, whiskey wouldn't be taxed again until 1862. In that year, President Abraham Lincoln was forced to reintroduce a whiskey tax to help finance the war. Once again, just as in the Revolutionary War and the War of 1812, booze was used to financially fuel warfare.

JOHN H. BEAM (1839-1915)

In 1779, the "Corn Patch and Cabin Rights" law was enacted by the Virginia General Assembly. It allowed settlers to claim 400 acres of land if they could prove they had built a cabin and planted a corn crop before January 1, 1778. The Kentucky soil and the area's unique water composition made conditions perfect for growing corn. And where corn grows, whiskey follows. Not only was corn a relatively easy grain to cultivate in fertile Kentucky fields, but the limestone and iron-rich Kentucky water produced a distinct flavor of whiskey.

The spirit Bourbon was first distilled in the late 1700s in these territories. Bourbon is a distilled spirit made from a fermented mash of corn that is aged in white oak barrels for at least two years.

Barrels of the now-famous Beam bourbon were first sold, in 1795, by Jacob Beam, under the name Old Jake Beam. Jacob's son, David, and his wife Elizabeth, had eleven children, amongst them were John Henry (known as Jack) and Joseph B. Beam.

Figure 1: Bottle of Jim Beam Bourbon Whiskey

As a teenager, Jack worked in the family's Old Tub Distillery in Washington County. When his father passed away, Jack, while only 21, moved the remnants of Old Tub to Early Times Station and christened both his distillery and his bourbon whiskey "Early Times." Beam's distillery produced the straight bourbon brands Early Times, Jack Beam, and A.G. Nall. Early Times had a very high corn content, 77 to 79 per cent, and it quickly gained fame as a superior bourbon whiskey. By the 1870s, Early Times was one of the few nationally-distributed whiskies.

One of Jack's innovations was the introduction of the heated warehouse. Climate-controlled rackhouses, avoiding temperature fluctuations, allowed the whiskey to age at a more even, steady rate.

"He is a great believer in the early time methods of making whiskey, which includes mashing the grain in small tubs, and boiling the beer and whiskey in copper stills over open fires. What he doesn't know about distilling isn't worth knowing," said the *Nelson County Register* of Jack in 1896. (Figure 1)

STILLS FOR THE MANUFACTURE OF ALCOHOLIC SPIRITS, PATENT NO. 165,201
John H. Beam and Joseph B. Beam, Bardstown, Kentucky, July 6, 1875

The still carries out the process known as the "sour and sweet mash," with four chambers communicating with each other by side-valves and central heaters. The still produces very pure spirits, with several evaporations and condensations occurring within the body of the still. The heat from the condensed vapors is reused to heat up the fresh mash or beer.

Twice as much mash can be distilled with the same amount of fuel that other stills use, in half the time used for charging cold mash. The closed charging-vessel prevents the loss of vapor while the mash is heating. Impurities are washed out and condensed in the chambers, with higher yields because the mash is charged while boiling.

CHARLES FLEISCHMANN (1834-1897)

On November 3, 1834, near Budapest, Hungary, Alois and Babette Fleischmann welcomed their son Charles into the world. Charles was educated in Vienna and Prague before emigrating to America in 1866. After working in a distillery in New York for two years, Charles and his brother, Maximilian (Max) moved to Cincinnati, OH, where the two of them, along with the distiller James Graff, founded the Fleischmann Company "to manufacture compressed yeast and distilled spirits."

Charles Fleischmann patented many of his inventions between 1869 and 1888. They included an improved distilling apparatus, a new process for aging liquors, an improved cotton gin, and a process for extracting oil from cotton seed. The patents also made various important improvements to sewing machines, machine cranks, and motors. Soon after beginning the business in Cincinnati,

Figure 2: Portrait of Charles Fleischmann

Charles began to manufacture compressed yeast, a more convenient form than the liquid yeast in use at the time. Fleischmann compressed yeast became very popular and soon overshadowed the rest of the business. As yeast must be used fresh, the company also maintained a fleet of wagons for the daily delivery of yeast. With this manufacturing and transportation infrastructure, Fleischmann's yeast quickly grew into an internationally-recognized brand, even showcased at the 1876 Centennial Exhibition.

In his later years, Fleischmann became involved in public service and even served as an Ohio state senator. (Figure 2)

APPARATUS FOR RECTIFYING ALCOHOLIC LIQUORS, PATENT NO. 103,320
Charles Louis Fleischmann, Cincinnati, Ohio, May 24, 1870

An apparatus for passing alcoholic liquors through compressed charcoal, while keeping the charcoal in a compact state by means of self-regulating pressure. The apparatus also revives the charcoal's rectifying powers by forcing air or other suitable gases through it when the charcoal becomes saturated with volatile substances.

The new mode of rectifying requires a smaller amount of coal, revivifying it within its portable and easily cleaned apparatus. The method minimizes loss by evaporation, and uses little alcohol.

TOBACCO

Tobacco is a native plant to North and South America, and historically, it has been one of the most important crops grown by American farmers. It is also one of the most widely-used, addictive substances in the world. Tobacco finds its roots in the same family as the potato, pepper, and belladonna (AKA "deadly nightshade"). Tobacco was also once believed to be a cure-all and had sacred associations amongst Native Americans.

The United States' long association with tobacco reaches as far back as 1492 when Christopher Columbus was offered dried tobacco leaves as a gift from Native Americans. John Rolfe introduced the tobacco plant to the Virginia colony at Jamestown. In 1613, Rolfe (who a year later would marry Indian princess, Pocahantas) grew a mild variety of tobacco from seed brought from the West Indies. Rolfe shipped a few barrels of his product to England, and soon, tobacco quickly became popular in Europe, mostly in the form of snuff and pipe tobacco.

Tobacco in America was initially viewed as a temporary crop, until settlers could plant something else. Tobacco commanded very low prices. In Virginia, during the 17th Century, tobacco could be bought for pennies per pound.

Tobacco has the unfortunate drawback of quickly draining the soil of its nutrients. As a result the land provided only three fruitful growing seasons. To become productive again it had to lie fallow for three additional years before re-planting. This created an accelerating demand for fresh farmland.

Tobacco cultivation is also labor-intensive, so an expanding workforce was needed. Indentured servants were the first to meet the need, followed by African slavery. Amidst its grim and immoral efficiencies, costs came down and profits went up. Small tobacco farms quickly became large.

Tobacco became the colonies' first "killer app". It became so valuable as a commodity that it was once used as money in several American colonies. In 1776, tobacco was used to finance the Revolutionary War — it was used as collateral for loans from France. The Civil War boosted the consumption of tobacco in the portable forms of cigars, followed by the then relatively new cigarettes. In the way of these new smoking technologies, chewing tobacco declined after 1890. Until the 1960s, the United States not only grew and processed, but exported, more tobacco than any other country in the world.

Originally, the tobacco industry produced snuff and loose tobacco for pipes, then came cigars and chewing tobacco. Eventually, the slender, more portable design of the cigarette arrived and quickly gained in popularity. Before the manufacture of cigarettes, plug (chewing tobacco) was the most popular form of tobacco consumption.

In 1870, Albert Pease of Dayton, OH invented a machine that chopped up tobacco for cigarettes, which, up to that time, had been exclusively made by hand. James Bonsack created a machine that first chopped the tobacco before dropping a measured amount of the cured leaf into a long paper tube. The machine then rolled and pushed the tube by a slicer which cut it into individual cigarettes. This machine operated at thirteen times the speed of a human's ability to roll cigarettes. Bonsack's invention revolutionized cigarette production and sales. For instance, The American Tobacco Co. (a consortium of five tobacco companies), listed $3 million in sales on the American Stock Exchange in 1890. With the significant help of machine-rolling technology, that number increased to $316 million by 1903.

GEORGE S. MYERS (1832-1910)

In 1878, in St. Louis, MO, George S. Myers went into partnership with John Edmund Liggett. Liggett & Myers Tobacco Co. (L & M) was born. At the time, St. Louis, MO and Durham, NC were key centers of tobacco production. In 1876, Liggett & Myers introduced L & M plug chewing tobacco. By the late 1880s, the company had begun producing cigarettes, too, and by 1885, they had become the world's largest manufacturer of plug tobacco.

John E. Liggett died in 1897. Two years later, Liggett & Myers was acquired by the American Tobacco Company. George S. Myers died in 1910 and the following year, the U.S. Court of Appeals issued a Dissolution Decree to the American Tobacco Co., creating the opportunity for Liggett & Myers Tobacco Co. to be reborn. It was, and was once again headquartered in St. Louis. (Figure 3)

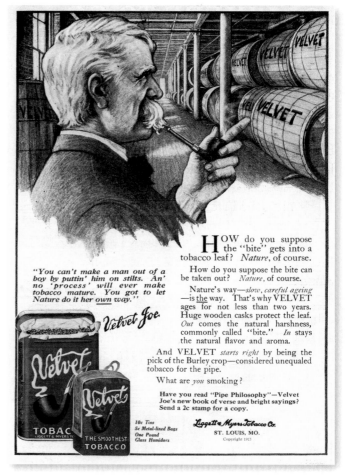

Figure 3: Advertisement for Liggett and Myers Tobacco Co.

PLUG-TOBACCO MACHINE, PATENT NO. 215,473
George S. Myers, St. Louis, Missouri, May 20, 1879

An apparatus that improves the way plug tobacco is compressed into rolls and then cut into plugs. The plug-cutting knife cuts vertically without interfering with the strip as it comes from the rolls.

RUDOLPH FINZER (1845-1919)

By the late 1800s, Louisville, KY had a booming business revolving around tobacco. The city boasted fifteen warehouses, sixteen tobacco manufacturing plants, and seventy-nine smaller firms producing cigars and snuff.

One of Louisville's more prominent plug tobacco makers was Five Brothers Tobacco Works, started in 1866 by the five Finzer Brothers: John, Benjamin, Frederick, Rudolph, and Nicholas. Born in Canton Berne, Switzerland, they came to the U.S. as boys.

By 1887, the Five Brothers company was producing four million pounds of plug tobacco and one million pounds of smoking tobacco a year. At its peak, the company is said to have had 400 workers. The Five Brothers factory was destroyed by fire in 1880, but was quickly rebuilt on the same spot. The company also published a trade paper, The Tobacconist, with a monthly circulation of 32,000. In 1904, the company fell to the juggernaut combine of the American Tobacco Co., and by 1924, it was out of existence. (Figure 4)

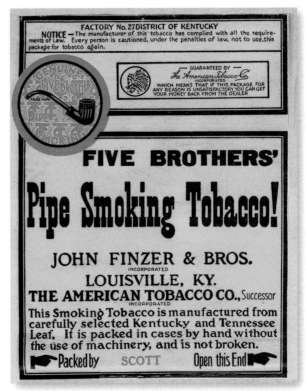

Figure 4: Advertisement for John Finzer and Brothers Tobacco

TOBACCO-CADDY, PATENT NO. 288,182
Rudolph Finzer, Louisville, Kentucky, June 1, 1880

A box (caddy) for packing plug-tobacco, allowing irregular or extra numbers of plugs to be packed in ordinary sized caddies. Tobacco to be shipped or stored properly must be tightly packed in compact solid packages, to keep out air and moisture and to prevent drying or molding. It is packed in caddies in even layers, making a solid compact mass, which is then pressed under pressure until all the air and moisture are removed. The lid of the caddy is then pressed in against the tobacco until it becomes air-tight. These caddies are usually made to contain a fixed number of pounds. If an irregular number of pounds or plugs are ordered they cannot be properly and tightly packed in these regular sized caddies. The Finzer caddy consists in making one or both heads grooved on the inner side to receive one or more plugs fitted snugly in the groove, so that when the heads are pressed it makes a solid compact mass with no unfilled space, thus being able to pack one or more extra plugs or pounds in the ordinary regular sized caddies.

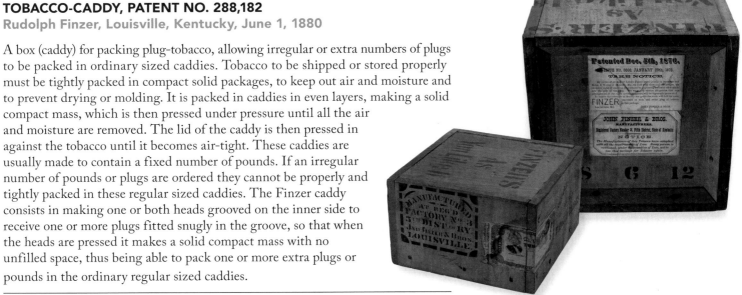

FIREARMS

For most of human history, warfare has been a battle of brawn before brain. But the invention of the firearm changed all that. Suddenly, the weakest army could be equal to the strongest—if supported by similar or superior firepower.

The cannon, bombard, mortar, musket, pistol, and petard all extend back through the pages of history. Like gunpowder, the cannon, the most ancient of firearms, is believed to have originated in China, around 950 AD. The projectiles used by early cannons were balls of stone, some enormous in size. The breech-loading feature of the cannon was introduced early in its history although it did not become a military mainstay until late 19th century. The rifling of the bore, with spiraling grooves, brought greater efficiency to the weapon, the techniques' rotating effect causing projectiles to maintain a truer flight.

The first cannons came to the Americas with Columbus, but it wouldn't be until the 19th century that improvements in accuracy, range, rapidity of fire, and the penetrating power of projectiles allowed them to dramatically influence the outcome of combat.

During the Civil War, a new and terrifying weapon made its debut, a weapon though small in caliber, that rivaled the cannon in its lethality: the machine gun. With this device, metal balls of comparatively small size were discharged uninterruptedly and in rapid succession. Of these new weapons, the first machine gun to bring controlled, sequential automatic fire into combat was the Gatling gun, invented in 1861 by Dr. R.J. Gatling.

Firearms have always relied on the burning of a propellant to provide them with high velocity projectiles. Black powder was originally used as that propellant, a mixture of saltpeter (potassium nitrate), charcoal, and sulphur. Black powder was used in all firearms until smokeless and other forms of propellant were invented in the late 1800s.

Modern firearms use smokeless powder and other explosive compounds such as cordite.

The 19th century proved to be the defining moment for handguns, when Samuel Colt invented one of the most ingenious weapons ever designed. Colt designed the firing process around a revolving cylinder, with several firing chambers, and one cartridge per chamber.

Three of the most important Americans associated with guns were Samuel Colt, inventor of the modern revolving pistol, Oliver Winchester, founder of the Winchester Repeating Arms Co., and John Browning, one of the greatest gun designers in history, founder of Browning Arms. (Browning is also known for the lever-action rifle, but he only developed it, he did not invent that technology.)

It was Oliver Winchester's Model 1873 lever action-rifle that earned the title, "the gun that won the West." The name Winchester has become synonymous with long guns, just as Colt is for handguns.

Samuel Colt not only made the first revolving pistol (the famous Colt Revolver), he also improved Eli Whitney's system of interchangeable parts and took it to its logical conclusion by the development of the assembly line. Colt began manufacturing his revolver in Paterson, NJ, in 1836. The Mexican War brought a flood of orders for Colt. He got Whitney's son to first make the revolvers for him and later he built his own factory.

These three men and the companies they founded are responsible for the development of virtually every important type of firearm except the double-barreled shotgun and the bolt-action rifle, both European in invention.

Of the three, let's take a longer look at Eli Whitney, and his son, Eli Whitney, Jr, and one of junior's critical inventions, breech-loading.

ELI WHITNEY (1765-1825) AND ELI WHITNEY, JR. (1820-1894)

Eli Whitney personified Yankee ingenuity and inventiveness. Whitney, forever famous as the inventor of the cotton gin, was born in Westborough, MA, on December 8, 1765. His father, for whom he was named, was a farmer. In 1789, Whitney left his Westborough home for Yale in New Haven, CT. At 28, he completed his college studies before heading south in hopes of putting his education to work. Five years later, in 1794, Whitney revolutionized the cotton industry with his invention for removing seeds from cotton fiber.

Patented in March of 1794, Whitney's cotton gin (Patent No. 72X) transformed the United States' economy, turning cotton from a luxury item into something that anyone could afford. A worker could turn Whitney's cotton gin by hand and produce 10 times as much cotton as hand-picking it. One man with a gin driven by a horse or water power could produce more than 50 times what a hand-picker could.

American industry also took a great leap forward when, in 1798, Eli Whitney began making rifles with interchangeable parts in his factory near New Haven, CT. The town of Whitneyville grew up around the plant to house its workers. Whitney proved to be one of the most important figures during the Industrial Revolution, a forefather of mass production. Before Whitney, a "manufactory" was a workshop where each artisan made each of the parts of a product by hand and then assembled them into the finished whole. Whitney concentrated not on the product itself, but on the tools required to make it. It was his aim to insure that "the tools themselves shall fashion the work and give to every part a just proportion," i.e., precision parts manufacturing. Whitney also became the first person to make and use power-driven tools.

The military wanted to be able to repair guns on the

Figure 5: Portrait of Eli Whitney

battlefield by swapping out their parts. Up until then, every rifle had been made by hand, from stock to barrel, and the parts of one gun did not fit another. Whitney proposed making all of the parts so identical that they could be interchangeable.

On June 21, 1798, Eli Whitney received a contract from the U.S. government to manufacture 10,000 muskets, at $13.40 each. In perhaps a fit of hubris, he promised to deliver all 10,000 units by September 30, 1800, with the first 4,000 to be produced in only 15 months. In the end, it took him nine years to complete the rifles he'd promised to deliver in two.

Thomas Jefferson wrote of Whitney, "He has invented molds and machines for making all the pieces of his locks so exactly equal, that take 100 locks to pieces and mingle their parts and the hundred locks may be put together as well by taking the first piece that comes to hand. This is of importance in repairing, because out of 10 locks disabled for want of different pieces, 9 good locks may be put together with out employing a smith."

Whitney's machine-made muskets were deemed superior to any imported or hand-made at home. However, everything was not as interchangeable as it seemed. It was later discovered that Whitney had secretly marked the parts that were to go together. In 1966, the Roman Numeral VI was discovered on the inside lock of a Whitney musket in the gun museum in New Haven.

Whitney strongly believed and advocated for the mechanization and interchangeability of parts. His ideas became the foundation for the machine tool industry and mass production. Samuel Colt gave the Whitney factory his order for 1,000 Walker pistols in 1847, and in 1858, Oliver Winchester began using the facility to make Winchester rifles.

Eli Whitney's son, also named Eli, was born in 1820 and was five years old at the time of his father's death in 1825.

When his father passed away, he too became known simply as Eli Whitney (not junior). In 1841, son Eli took over the operation of the armory from his cousins, the Blake brothers. Eli junior was no stranger to invention. He was a leader in the development of gun design and manufacture. He received many patents over a period of 30 years and was one of the first manufacturers of percussion-cap guns. Son Eli Whitney died in 1894 at the age of 74.

The Eli Whitney family of New Haven, CT produced multiple generations of inventors whose creativity had a profound impact on the American way of life. (Figure 5)

BREECH-LOADING FIRE ARMS, PATENT NO. 112,997
Eli Whitney, New Haven, Connecticut, March 21, 1871

The breech-block turns to the rear to open the barrel for inserting the charge. The mechanism, which is combined with the hammer and breech-block, yet independent of both, locks the breech-block in position.

Whitney's Claims for his invention:

1. In combination with the breech-block and hammer the cam or cams provided with thumb piece, operating upon the same center with the hammer, but independent of the same, to release and lock the breech-block.

2. The hammer may be set at half-cock by means of the cam.

3. The arrangement of the latch, to catch and hold the cams and be tripped by the movement of the breech-block.

4. The combination of the hammer with the cams and the latch to couple and hold the hammer together on the cock-notch.

CHARLES BROWN (1796-1865)

In the early 1800s, whaling fever overtook Warren, RI. At one time, there were as many as 26 whalers mooring on its docks. Maritime activities were very important to the city's economy, and its waterfront was alive with foreign vessels and ships setting sail for far off lands or long whaling voyages.

This was backdrop for the whaling vessel Rosalie setting sail, in 1821, with John Gardner as her captain and Charles F. Brown as his chief mate. On one voyage, they allegedly killed 101 whales, 49 of them by Chief Mate Brown himself. Thanks to his many successful whaling adventures, Brown was eventually made captain of the Rosalie, in 1828.

Utilizing his knowledge of the seas and experience as a whaling captain, Brown began submitting numerous patent applications. He received many for ideas related to navigation, along with patents for harpoons, a steering rudder, telescopic mast yards, and an adjustable screw propeller. He also made use of his knowledge of artillery, receiving patents for improvements to artillery ordnance mounting, a wheel for gun carriages, and a firearm breech-loading ordnance.

Charles Brown not only received patents in the U.S., but also in France and England. His English patent for Improvements in Ordnance and Firearms was registered under the name of William Edward Newton as England would not issue a patent to a U.S. citizen. (Figure 6)

Figure 6: English Patent assigned to Charles F. Brown for Ordnance and Fire Arms

FIRE ARMS, PATENT NO. 30,045
Charles F. Brown, Warren, Rhode Island, September 18,1860

A cannon that fired repeatedly as it was drawn over the ground, effective in advancing to meet, pursue, or retire from an enemy.

Brown not only patented his invention in the U.S. but also, in 1860, he obtained patents from France and the United Kingdom for his cannon.

MODE OF MOUNTING ORDNANCE, PATENT NO. 13,249
Charles F. Brown, Warren, Rhode Island, July, 17, 1855

A method of mounting cannons and other ordnance in ships, forts, land, and floating batteries that closes, as nearly possible, the ports or embrasures all around the piece, to prevent the entrance of an enemy's shot, and the smoke of firing from affecting the gunners and their attendants.

WHISKEY STILL, PATENT NO. 79,373
John G. Mattingly and Benjamin F. Mattingly, Louisville, Kentucky, June 30, 1868

An "Improvement in Boilers for the Purpose of Distilling Whiskey or other Spirits" that kept beer from burning or encrusting the boiler during distillation.

APPARATUS FOR AGING AND PURIFYING LIQUORS, PATENT NO. 221,316
Charles H. Jacob and John W. Lochner, Cincinnati, Ohio, November 4, 1879

An adaptation for rapidly and economically imparting the quality usually imparted by aging. The apparatus increases evaporative and cooling surfaces but prevents the escape of essential oils.

CIGAR MACHINE, PATENT NO. 131,474
George W. Tanner, Providence, Rhode Island, September 17, 1872, Assignor to Tanner Cigar Machine Company

A novel combination and arrangement of bed-rollers with two releasing rollers that perform the functions of molder, binder, and wrapper to make perfect cigars, requiring only that the tuck ends be cut or squared off.

SPRING-FOLLOWER FOR CIGAR BOXES, PATENT NO. 311,597
Moses Michaelis, New York, New York, February 3, 1885

The innovation prevents cigars from being thrown about in the box and injured when they are packed away in a trunk. The device is placed in the top part of the cigar box, and the followers adapt themselves to the heights of the different layers of cigars to keep the cigars in place.

CIGAR BOXES, PATENT NO. 159,825
Sigmund Jacoby, New York, New York, February 16, 1875

A metallic cigar box with a hole for branding. The brand required by the revenue law of the time could be applied through the hole and remain in view for inspection.

CIGAR LIGHTERS, PATENT NO. 206,577
James M. Keep, Jersey City, New Jersey, July 30, 1876

The lighter's hammer and actuating-spring device are separate and easily removable, without removing the fastening screws, unlike other lighters of the time.

DEVICE FOR MARKING PLUG TOBACCO, PATENT NO. 204,042
Harry C. Holbrook, Louisville, Kentucky, May 21, 1878

A device for forming and marking plug-tobacco, using a series of raised panels and alternating depressions. The raised panels have sunken letters, marks, or symbols, which press into the tobacco-plug as it is formed.

FINE-CUT TOBACCO MACHINE, PATENT NO. 204,539
George B.F. Cooper, New Albany, Indiana, June 4, 1878

A machine for dressing or separating the shorts (refuse) from the long-cut fibers of saleable tobacco, and also drying the tobacco while dressing either by dry or steam heat.

SHELL–PROJECTILE, PATENT NO. 189,358
Benjamin B. Hotchkiss, New York, New York, April 10, 1877

Previous shells and explosive projectiles sustained the firing charge without breaking into pieces upon impact. This shell's walls break into a large number of fragments under the force of the exploding charge.

Hotchkiss (1826-1885) was one of the leading American ordnance engineers of his day. In 1867 he developed a revolving barrel machine gun known as the "Hotchkiss Gun."

CARTRIDGE-LOADING DEVICE, PATENT NO. 167,778
Warren Noyes, Gorham, New Hampshire, September, 14, 1875

A convenient set of devices for charging the cartridge cases used in sporting breech-loading firearms.

PERCUSSION-PROJECTILE, PATENT NO. 13,469
Augustus McBurth, Elizabeth, New Jersey, August 21, 1855

A bomb shell like a cannon ball, more destructive and accurate than the current bomb shell. Its "four arms and eight flutes with sharp edges enable it to pass through the air with less resistance and greater impetus than any other ball or bomb shell ever used."

EXPLOSIVE COMPOUND, PATENT NO. 288,516
Harry D. Van Campen, Belmont, New York, November 13, 1883

This new and improved blasting compound is safe to handle, does not accidentally explode and leaves no noxious fumes behind it after explosion. The compound contains pulverized tan bark (50 parts), dextrine (10 parts), cryolite (5 parts), potassium nitrate, (15 parts) and nitro-glycerine (20 parts).

BLASTING CARTRIDGES, PATENT NO. 152,053
Julius H. Striedinger, New York, New York, June 16, 1874

A cartridge that facilitates the destruction of buildings in populated cities. When placed against the walls of the building and exploded, it "secures the desired result without danger to other buildings."

FUSE LIGHTERS, PATENT NO. 180,267
John W. Platt, Mineral City, Nevada, July, 25, 1876

Lighting blast fuses by placing a short piece of candle under the fuse often fails. This fuse lighter attaches directly to the fuse. It cannot become detached after it is lighted, or burn up without igniting the fuse of the blasting material.

METHOD OF SETTING HAIR-TRIGGERS OF RIFLES, PATENT NO. 181,855
George O. Leonard, Red Bluff, California, September 5, 1876

Usually, the set-trigger was set by throwing the trigger forward with the thumb, which took time and effort. Leonard's improvement, a set screw or cam in the finger lever of a rifle that engaged with the back of the trigger, set the set-trigger whenever the finger lever moved up or down.

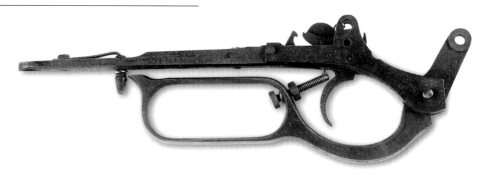

CHAPTER 4

Heat, Light & Fire

"Education is not the filling of a pail, but the lighting of a fire."

—William Butler Yeats

Fire has always satisfied the human need for warmth, food preparation, and light. Creating fire by rubbing one substance against another stretches back to the dawn of Man. As people learned how to make and maintain fire, they began to explore ways of improving it as a light source. In ancient caves, scientists have discovered shelves dug into rock walls where fires were maintained for lighting the space. People began to differentiate between materials that burned brightly and lasted longer, and those that were more suitable for cooking. Certain wood, such as pine, was found to burn steadily without much smoke. Better light was obtained by soaking the wood in oils and waxes from animals or berries. Meat grease gave a bright light, and the fats from birds and fishes were used in torches.

For centuries, the candle was a major source of illumination. Lamps used vegetable oils and greases, but even the best lamps had problems; they were smoky, smelled bad because of the rankness of the fuel, and incomplete combustion threw off smoke and soot. Many inventors spent their time and skill in ornamenting the lighting vessel and making artistic and beautiful patterns; shades, reflectors, and shields of various kinds were designed for both lamps and candles. But these innovations left unsolved the problem of proper combustion of the lamp fuel. Various dangerous fuels, such as camphene (a mixture of turpentine and alcohol), were common while the oil-burning lamp was being perfected. Explosion was a real danger. Because of these dangers, many families continued to use candles, as they were safer and more dependable.

The introduction of gas lighting was revolutionary, but as it developed, so did the ultimately even more revolutionary technology, electricity. Two giants in this burgeoning new field were Benjamin Franklin and Thomas Edison.

BENJAMIN FRANKLIN (1706-1790)

Bifocals, lightning rods, the Franklin stove: these disparate inventions all came from the fertile mind of Benjamin Franklin. Yet Franklin refused to patent any of his ideas. "As we benefit from the inventions of others, we should be glad to share our own," he said. Called a "one man Renaissance," Benjamin Franklin developed the theory of positive and negative electricity, and performed his famous kite experiment in 1752, drawing down electricity from the clouds and charging a Leyden jar (a kind of early capacitor) from a key at the end of a string.

Franklin spent a great deal of time trying to find a cure for smoky chimneys and developed a reputation as a heating expert. In the early 1700s, the open fireplace was the most common method of heating a home and cooking. But most of the heat went up the chimney. The Franklin stove, invented in 1742, was a metal-lined fireplace that stood in the middle of a room, with rear baffles for improved air flow. By pulling the stove away from the wall, Franklin increased its heating efficiency, with the stove's flue serving as a simple radiator. The Franklin stove was more efficient and, significantly, allowed people to warm their homes more safely.

Franklin also invented the lightning rod, which protected buildings and ships from lightning strikes and damage. The metal rod at the top of a building is grounded through a wire, and can conduct the electrical charge from a lightning strike harmlessly into the earth.

THOMAS ALVA EDISON (1847-1931)

Thomas Alva Edison was the first to supply electricity commercially, and is responsible for the modern system of generating current in a central station and distributing it out to homes and factories. But Edison is perhaps best remembered for his filament of carbonized thread sealed in a vacuum — the first practical incandescent electric lamp. He patented his invention in 1880, then devised an electric generating system, which by 1882, was in use in his New York city power plant.

A prolific inventor, Edison received 1,093 U.S. patents, along with thousands of additional ones from dozens of other countries — a number no one else has ever reached. Two of his most important were Patent #223,898 for the Electric Lamp (Light Bulb), patented on January 27, 1880, and Patent #200,521 for the Phonograph or Speaking Machine, patented on February 19, 1878. Among his other well-known patents were the motion picture, the telegraph, and the telephone.

Born in Milan, OH, Edison was the youngest of seven children. He only went to school for three months, his teacher dismissed his unusual mind as being "addled". His mother, a teacher, then began to supervise his education at home.

In 1876, Edison set up an industrial research laboratory in Menlo Park, NJ, the first of its kind. He made it his goal to produce a new invention every 10 days, and during one four-year period he obtained an average of one new patent every five days, earning him the nickname "The Wizard of Menlo Park." Thomas Edison credited his success to hard work, famously saying, "Genius is one percent inspiration and ninety-nine percent perspiration."

SIR HIRAM STEVENS MAXIM (1840-1916)

Sir Hiram Stevens Maxim was born in Sangersville, ME, in 1840, the eldest son of a mechanic. While born American, he would later become a British citizen. He is best known for his invention of the rapid-fire machine gun. Maxim had a genius for invention and is undoubtedly one of the most brilliant inventors of his day. And he managed all this on only five years of formal education in a backwoods school house. Maxim's ingenuity surfaced early, when he invented an automatic mousetrap. Unlike most mousetraps of the time, which could only fit one mouse, his had room for a number of them and the trap would re-set itself each time. As was the case with many of Maxim's other early inventions, he never took out a patent for the device. In his youth, he wandered the eastern United States and Canada, taking odd jobs as a carriage painter, cabinet maker, and mechanic while still finding time to teach himself science and engineering. Maxim's innate mechanical genius combined with a natural ability to draw. He was an avid reader, given to experiments, and was always at work refining and perfecting his inventions.

At 26, Maxim obtained his first patent for a hair curling iron. He also invented the first automatic sprinkler system. At the same time that it sprayed water to put out the fire, it would telegraph the police and fire station with the fire's location. Far ahead of his time, he had difficulty selling his invention because it was believed too good to be true. Seventeen years later, the first sprinkler system was installed in a Massachusetts cotton factory. By that time, his patent had expired. This would not be Maxim's only instance of being ahead of his contemporaries.

Maxim was also known for being absent-minded; so much so that he had stickers printed up that read:

THIS WAS LOST BY A DAMNED FOOL NAMED HIRAM STEVENS MAXIM WHO LIVES AT 325 UNION STREET, BROOKLYN. A SUITABLE REWARD WILL BE PAID UPON ITS RETURN.

It was in Brooklyn, NY, that Maxim developed machines for generating and illuminating gas, establishing the Maxim Gas Company. In 1878, Maxim became chief engineer of the United States Electric Company, Edison's rival in the field of electricity. In 1881, the United States Electric Lighting Company sent Maxim to Europe to represent the company at the Paris Exhibition. Here he was awarded the Cross of the Legion of Honour for his electric pressure regulator. Maxim first worked with gas illumination, then electricity. He is credited with introducing carbon filaments for electric light bulbs, and narrowly missed being the acknowledged inventor of the incandescent lamp. In a patent suit, Thomas A. Edison proved priority by only a number of days. Maxim's company produced his incandescent lamp, which was praised by some experts as the best of its time. In 1883, it was reported that more than 50,000 Maxim incandescent lamps were being used in the United States. However, there were problems with both the British and American electrical patents and both soon became common property.

Maxim received his greatest recognition for inventing the world's first fully automatic portable machine gun in 1884 in London. Known throughout the world as the Maxim gun, this invention revolutionized warfare. His gun could fire 500 rounds per minute, equal to the firepower of a hundred rifles. Maxim's machine gun used a simple concept. Maxim used the energy of each bullet's recoil force to eject the spent cartridge and insert the next round. His machine gun would fire until the entire belt of bullets was spent, and no external source of energy was required. He later invented a smokeless cartridge using cordite, which improved the effectiveness of his machine gun. He offered his weapon to the United States War and Navy Departments, both of which declined it on the ground that it was impractical — little more than an interesting and ingenious mechanical curiosity. The United States' lack of interest hurt him deeply, and he took his gun

Figure 1: Portrait of Sir Hiram Stevens Maxim

to England where it was accepted by the British War Office. The following year, Maxim sold his gun to most of the major powers in Europe. For the next 30 years, Maxim lived outside the U.S. and returned to the States only to visit. Maxim became a national hero in England, and in October 1899, he became a British subject. Within a year he was knighted by Queen Victoria, becoming Sir Hiram Maxim, in recognition of the contribution his gun had played in the success of the British Army.

Also a pioneer of flight, Maxim was very aware of the great significance that the flying machine would have. Several years before the Wright brothers learned to fly, Maxim's machine rose briefly in the air but was unable to achieve sustained flight. In 1894, in England, he launched his machine with two passengers and himself on board. Maxim's plane weighed 7,000 pounds, had a wing area of 4,000 square feet, and was driven by two 180-horsepower steam engines. The flying machine rose only a few inches, while running along a guide rail. An outrigger broke through the rail and Maxim cut the power to avert a crash. It was the first time a heavier-than-air machine had ever lifted off the ground under its own power. In his later years, Maxim continued to invent things. In 1907, he invented a steam-powered vacuum cleaner. It was compact and could be moved around in domestic use. Also, inspired by his own experiences with chronic bronchitis, he invented an improved medical inhaler. In 1912, Maxim anticipated the theory of radar with the idea of a mechanical means of detecting unseen objects, such as ships at sea or icebergs in bad visibility. It could also give the distance and size of the unseen object. His method was the same as would later be used in radar and sonar. In 1916, the last year of his life, he took out two provisional patents for gun silencers and two for breaking down crude oil into lighter hydrocarbons such as petrol and related flammable products.

Throughout his life, Sir Hiram Maxim epitomized

the spirt of inventiveness. Winston Churchill, Thomas A. Edison, Orville and Wilbur Wright — not only were these men contemporaries of Maxim's, they were, in the cases of Edison and the Wright brothers, rival inventors. Edison, Maxim's greatest rival in the early years of electricity, called him "the most versatile man in America." The young Winston Churchill and Maxim were both members of the Aeronautical Society in England. Churchill had great respect for Maxim and wrote letters on behalf of him, promoting Maxim's knowledge of aircraft.

Maxim liked to refer to himself as a "chronic inventor." He held 122 U. S. and 149 British patents. The scope and breadth of Maxim's inventions throughout his lifetime attest to the man's extraordinary mind. (Figure 1)

CARBURETOR, PATENT NO. 212,857
Hiram S. Maxim, Brooklyn, New York, March 4, 1879

An automated machine for carbureting air for illuminating purposes, converting gasoline into a vapor under heat and pressure, and injecting sufficient air into its escaping force to reduce it to the proper density for illuminating or mechanical purposes.
"The apparatus employs a gas-holder, a pump for injecting the gasoline into the vaporizing portion of the machine, and a weight that is raised by the gas-holder, and remains supported until the gas-holder descends and liberates the weight, and it falls to actuate the pump."

Figure 2: E. and T. Fairbanks Company Factory

THADDEUS FAIRBANKS (1796-1886)

Fairbanks Standard Scales is one of the nation's oldest industrial manufacturing companies. The company was started by two brothers, Thaddeus and Erastus Fairbanks. Thaddeus was a mechanic and builder, a wagon maker by profession. He had many ideas for inventions and built a foundry in 1823 to manufacture two of his inventions—the cast iron plow and a stove. His brother Erastus joined the business in 1824, and brother Joseph joined after their first successes. The two brothers formed the E. & T. Fairbanks Company in St. Johnsbury, VT. While in business together, they realized that current weighing scales were inherently inaccurate, so Thaddeus decided to invent a more dependable weighing machine. Through an arrangement of levers, he reduced the weight needed to counter-balance a load. He also dug a pit for the levers, placing the platform level with the ground, thus entirely doing away with the task of having to hoist up the entire load. In his first design, Thaddeus rested a platform on two long levers which were connected to a steelyard, upon which the counter-balance was placed. Although he achieved accurate weights, Thaddeus was not satisfied with the stability of the design. He added two short levers to his long ones, establishing support points at all four corners of the platform. In 1830, Thaddeus built his first professional quality scale and applied for a patent.

The Fairbanks brothers' scales were sold throughout Europe. In 1846, the company began trade with China, and two years later, began selling to Cuba. By the time of the Civil War, Fairbanks' scales were known throughout the world, from the Caribbean to South America, India to Russia. By 1865, the company was producing 4,000 scales a month. Fairbanks scales were known for their accuracy, dependability, and longevity.

By 1882, the company was producing more than 80,000 scales annually. By 1897, they held 113 patents for inventions in weighing. Fairbanks produced 2,000 standard model scales, but made as many as 10,000 different models and custom systems. (Figure 2)

COOKING AND HEATING STOVE, PATENT NO. 8763X
Thaddeus Fairbanks, St.Johnsbury, Vermont, April 14,1835

Fairbanks' invention is for improvements in the mode of constructing stoves for cooking and for warming apartments. The stove is best adapted for the burning of wood and may be either square or oval having the general form of the common coal stove. This model was reproduced after the fire of December 1836 when the entire patent office was destroyed along with approximately 10,000 models. This is one of about 2,000 of the models that were reproduced by the original inventors. The X after the patent number signifies that it is prior to the numbering system that began on July 4, 1836.

HEATING STOVE, PATENT NO. 164,899
Jabez K. Babcock, Phelps, New York, June 29, 1875

An "Inversible Heating Stove", with a fire-pot composed of annular sections with vertical air heating flues or tubes. The fire-pot is mounted upon trunnions so it may be inverted.

GRATE FOR FURNACES, PATENT NO. 68,139
William A. Wilson, James Smith, Liverpool, England, August 27, 1867

Two English inventors improved the design of furnace grates, carrying the burnt-down fuel towards the flue, keeping the spaces for admitting air between the bars open, and preventing clinker formation. "Fuel is more perfectly consumed and a greater amount of heat given out than in ordinary furnaces, and the escape of black smoke into the atmosphere is prevented."

HAND LANTERN, PATENT NO. 56,543
Carl Engelskirchen, Buffalo, New York, July 24, 1866

The lantern's base is fastened to the glass globe with a spring band or clasp which also forms a part of the skeleton frame work of the lantern. The frame work is permanently attached to the chimney-cap.

FIRE KINDLER, PATENT NO. 175,858
George W. Eldridge, South Chatham, Massachusetts,
April 11, 1876

The Fire Kindler consists of a piece of wood, (pine preferred) a piece of
paper, and six to eight pine cones. The cones are placed on either side of
the wood, then saturated with a mixture of melted resin and kerosene. The paper is
then wrapped around the whole bundle, which burns freely and kindles the fire.

APPARATUS FOR CARBONIZING AIR FOR
ILLUMINATING PURPOSES, PATENT NO. 44,560
Warren A. Simonds, Boston, Massachusetts, October 4, 1864

A cylindrical reservoir that facilitates "the vaporization of volatile liquids when used
for carbonizing the air for illuminating purposes." The reservoir is filled with light and
volatile naphtha, and air is forced from the meter through the connecting pipes.

COOKING STOVE, PATENT NO. 4,749
William H. Allen, Wellsburg, Virginia,
James Slocum, Brownsville, Pennsylvania,
September 10, 1846

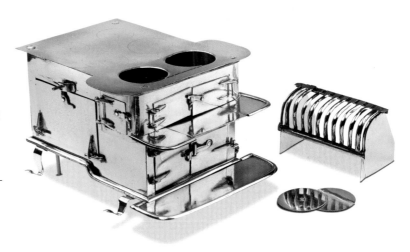

A cooking stove that converts from a wood-fueled
to a coal-burning stove by removing a movable
ash pit hearth, and placing a grate and plates in
the front part of the stove.

FIRE PROOF CURTAIN, PATENT NO. 186,657
William D. Baker, Rockland, Massachusetts, January 30, 1877

A fire proof curtain for theaters, with pockets sewn into it. When hung, it would be surrounded by pipes with stopcocks that when opened would fill the pockets with water. As the water soaked into the felt backing, it rendered the curtain fireproof and prevented fire from spreading into the theater.

In the 1860s and 1870s, theater stages were commonly lit by burning calcium oxide that would produce a bright lime color. The expression "in the limelight" came about from this process.

It was not unusual for sparks to get into the curtains, resulting in disastrous fires in theaters. If sparks got into entertainers' clothing, a wet blanket would be thrown upon them as they left the stage so as to prevent a possible fire, resulting in the expression "someone is a wet blanket."

ILLUMINATOR AND INCENSE BURNER, PATENT NO. 117,281
E. Warren Hastings
Boston, Massachusetts,
July 25, 1871

An illuminator "so constructed that the illuminating compound or powder used can be ignited instantaneously simply by the use of a movable lamp. When it is used as an incense burner the lamp is removed and the incense is placed in the cover of the urn."

CHAPTER 5

Locks

"I grew up in Brownsville; most of the kids I grew up with went to jail, not Yale. If they had heard of Yale, they thought it was a lock to pick."

—Mitch Leigh

The earliest safes in Colonial America were simply hardwood boxes bound with iron straps, with padlocks used to hold the lids closed. Some were soaked in brine and sheathed in iron for protection against fire. Breaking into such safes required no skill beyond the ability to swing a pickaxe or a sledge hammer. Robbers could simply smash a safe open or remove it through a window, though safe makers soon foiled this tactic by making iron safes that were too heavy to lift. Safes were used in offices and banks to protect against fires and burglars. Fire resistant safes designed to protect their contents were introduced in the early 1800s.

The work of two American inventors revolutionized the lock and safe industry throughout the world: Linus Yale Jr. and James Sargent.

LINUS YALE JR. (1821-1868)

Linus Yale Jr.'s most important invention was the Yale mortise cylinder lock. This key-operated mechanism was considered to be one of the most important developments in locks. It was a new type of mortise lock that added a pin tumbler cylinder to the side of the lock, sunk within the thickness of the door that housed the device. The lock's flat key (called a feather key) had notches cut to the proper depth which raised the tumbler pins to the exact height of the plug, allowing the plug to rotate freely. The special cam on the end of the plug operated the lock's dead bolt.

Many of the historically significant people covered in this book invented machinery that impacted a number of different disciplines. Linus Yale, for instance, invented a variety of tools. (See Chapter 17, Tools, for more information on Yale's life and other achievements.)

JAMES SARGENT (1824-1910)

Born in Chester, VT on December 1, 1824, James Sargent is best known for his invention of time locks for bank safes. His patented time lock mechanism is the grandfather of the time locks that banks use today.

In 1852, Sargent formed a partnership with his brother-in-law, Dan Foster, to manufacture apple parers. He joined the Yale Lock Company as a traveling salesman in 1862.

Sargent was a mechanical genius. When Yale introduced his new pin-tumbler lock, Sargent astounded his employer by picking it. Sargent invented an instrument he called a "micrometer" which, when attached to the dial of a lock, could reveal its combination by measuring the movement of its tumblers. Sargent then invented a combination lock that resisted his micrometer by including a large magnet. He called this the "Sargent Magnetic Bank Lock."

Sargent's greatest invention, the thief-proof lock, was developed in response to an epidemic of bank robberies in the latter half of the 19th century. Burglars would gain access to bank vaults by threatening bank employees into revealing the locks' combination. For example, on January 25, 1876, $1,250,000 was taken during the Northampton Bank Robbery in Northampton, MA when thieves kidnapped the cashier from his home and forced him to reveal the combination. It has been written that Sargent's invention was developed specifically to foil the notorious bank robbers Frank and Jesse James and their notorious gangs.

Sargent used two eight-day alarm clocks in his device called a chronometer lock or "time lock." The device only allowed the bolts on a vault door to release at a predetermined time. Once the lock was set and the door closed, not even the manufacturer of the vault could open the door before the allotted hour.

Sargent also invented the combination lock. His was the first successful key-changeable combination lock. In 1857, Sargent founded the James Sargent Lock Company to design and manufacture high security products. In just two years, 1860 and 1861, he invented the Micrometer, the Sargent Magnetic Bank Lock, the Sargent Automatic Bank Lock,

and the Sargent Time Lock. The last two inventions were accepted by the United States Treasury Department, and in 1869, Sargent won a federal contract to provide safe locks for the U.S. Treasury.

In 1864, Sargent moved to Rochester, NY and formed the Sargent & Greenleaf Co. with the financial aid of his former employer, Colonel Halbert Greenleaf. Now headquartered in Nicholasville, KY, Sargent & Greenleaf was acquired by Stanley Works in 2005 and is still considered the world's leading manufacturer of locks for safes, vaults, safety deposit boxes, and high-security devices.

The patent papers of James Sargent reveal his great role in the evolution of lock technology in America. Between 1865 and 1878, at least 17 known patents were issued to Sargent, and his locks are the basis for most combination locks manufactured today. (Figure 1)

Figure 1: Sargent and Greenleaf Model #2 Time Lock 1874

BOLTS FOR SAFE DOORS, PATENT NO. 85,246
James Sargent, Rochester, New York, December 22, 1868

An "improvement in locking the Bolt Work of Safe Doors." The bolt work may be set to open by the action of one lock or both.

The two keys may be held by different people and the locks may be set on different combinations. Then the bolt work cannot be opened till those two persons come together, preventing robbery by either key holder.

LOCK, PATENT NO. 213,252
James Sargent, Rochester, New York, March 11, 1879

The object of this invention is to improve that class of locks for doors, drawers, and other articles which may be set at any desired distance from the edge thereof, and also to construct a lock which shall be cheap, compact, and simple in construction, and readily applied. In this lock design, the bolt recess is placed in the body of the door or drawer to be locked. Leaving the wood on each side of the bolt strengthens the door where the lock is attached.

ELI WHITNEY BLAKE (1795-1886)

The first lock company in New Haven, CT was Blake Bros. Co., started in 1833 by three brothers, Eli Whitney Blake, Philos Blake, and John Blake, nephews of Eli Whitney, the inventor of the cotton gin. (See Chapter 3, Alcohol, Tobacco, and Firearms for information about Eli Whitney and his inventions.)

Eli Whitney (E.W.) Blake came to New Haven from Westborough, MA to attend Yale University. His uncle assumed the cost of his college education. Blake graduated from Yale in 1816 and went on to study law in Litchfield, CT. He returned to New Haven after receiving his law degree and became a lawyer but left the practice of law when his uncle, Eli Whitney, asked for Blake's help in running his gun factory in Whitneyville, CT. Blake made improvements in the machinery and the process of manufacturing arms. After Whitney's death in 1825, E.W. Blake and his brother Philos managed the business, and in 1836, they were joined by their brother, John.

The Blake brothers were all mechanically inclined and started their own business in the field of lock making. While still operating the Whitney Armory, they invented a new type of escutcheon door lock, and began to produce the door locks, latches, and other hardware that they invented. Their company was the first to introduce "mortise" locks and latches which are inserted in the the body of the door, superseding the previous clumsy and

Figure 2: Portrait of Eli Whitney Blake

disfiguring "box" locks and latches of English manufacture, which were affixed to the surface of the door. Their company was also a leader in the manufacture of carriage hardware.

In 1852, E.W. Blake was appointed to supervise the paving of the city streets in New Haven, CT. At the time, there were not a dozen miles of macadam road in all of New England. Blake became aware of the need for a machine to break stone, finding that the only method used to break stone into fragments was by hand-held hammers, costing two days worth of labor to produce only a cubic yard of road metal and this "… in coarser fragments than was desirable for a good road-bed." He wanted to develop something that could act on a considerable number of stones of different sizes and shapes at the same time, and then automatically remove the fragments when they reached the desired size. In 1858, he invented the first stone and rock crushing machine to provide material to build the nation's first paved roadways. Blake's solution was a pair of upright jaws converging downwards. The space between them at the top was large enough to receive the stones to be broken, and the space at the bottom was small enough to let the resulting fragments escape. The machine operated by "imparting to one of the jaws a short and powerful vibratory movement." This achievement earned Eli Whitney Blake the title, "Father of the American Road System." (Figures 2 and 3)

ESCUTCHEON LATCH, PATENT NO. 7945.5X
**Eli Whitney Blake, Philos Blake, John Blake,
New Haven, Connecticut, December 31, 1833**

A latch and bolt that substitute for a mortise latch and bolt. The latch's round shank receives a spiral spring that actuates the bolt. The bolt passes through a square hole in a plate on the edge of the door. The round hole through the door has its center exactly opposite to the center of the bolt, forming the chamber which receives the tumbler. The chamber is covered on each side of the door by an escutcheon plate.

Figure 3: Blake Stone Breaker c.1870

BURGLAR PROOF SAFES, PATENT NO. 115,221
Samuel S.B. Lewis, William H. Sterling, Troy, New York, May 23, 1871

A wedge-proof safe that has a bisected door, with each section's tongues and grooves interlocking with corresponding grooves and tongues in the door jambs.

The inventor claimed that with a safe constructed in this manner it would be impossible to use force or pry open the doors.

WINDOW LOCK, PATENT NO. 47,537
F.G. Ford, Washington, District of Columbia, May 2, 1856

A burglar-proof sash lock that secures the window sash from the inside, so they cannot be opened or raised from the outside. This is accomplished by the use of a face plate and double slotted tube, a sliding spring bolt and a double inclined plate catch. The lock system is operated by a removable key.

DOOR LOCK AND LATCH, PATENT NO. 256,665
Ole Flagstad, Hamar, Norway, April 18, 1882

A Norwegian design with an extremely complex mechanism which is detailed in three pages of abstract and three pages of drawings in the original patent papers.

This door lock is actually the equivalent of today's modern deadbolt locking system.

LOCK, PATENT NO. 281,713
Frank W. Mix, New Britain, Connecticut,
Assignor to the Corbin Lock Company, July 24, 1883

A cabinet lock with a hooked bolt for use on roller desks, chests, etc. It uses a flat key and is an economical and efficient lock to manufacture. Mix was a highly respected and accomplished machinist, toolmaker, and die maker in the lock industry in the New England states. He was employed by several major lock companies and served as superintendent for both the Eagle Lock Company and the Corbin Lock Company during his working career. It is thought that Mix took out more patents for trunk and cabinet locks than anyone else in the country. In 1870 he patented the Mix Lock, which the government adopted for mail bags.

COMBINED KNOB LATCH AND LOCK, PATENT NO. 73,498
Conrad Brown, Goshen, New York, January 21, 1868

A lock that uses a key to manipulate tumblers out of a notch in the door bolt.

BURGLAR PROOF SAFE, PATENT NO. 99,085
E.M. Hendrickson, Brooklyn, New York, January 25, 1870

A safe that cannot be blown open with explosives, because it has perforations or apertures that let the gas and pressure of the explosion out, dissipating its force and resulting in no harm or damage to the safe.

CHAPTER 6

Mining

"Diamonds are nothing more than chunks of coal that stuck to their jobs"

—Malcom Forbes

Gold was likely the earliest discovered metal because it is found in nature in its pure form. Other metals must be smelted, with rocks containing the metal needing to be crushed and then heated to separate the metal from the rock substrate. Further processing is often required, since ores frequently contain several metals, or the metal may contain impurities.

The first working metal was copper. At around 3000 B.C, in Mesopotamia, bronze was produced by melting it with tin. Indications of iron smelting were noted in Asia Minor about 1400 BC.

The first iron mining in the New World was begun in Virginia in about 1608. By the middle of the 18th century, iron ore had been discovered in all thirteen colonies.

Up until the Industrial Revolution, metal was smelted with charcoal. Early metalworkers learned that charcoal had a property that enabled it to change iron ore into iron.

Mining intensified as the need for metals increased. When armies needed weapons, the need for metal became increasingly urgent, and inventors focused on developing machinery that could get ore out of the ground faster and extract its metals most efficiently.

Humans began to mine petroleum in the 19th century. Before its advent in the U.S., candles or whale oil were burned. Whale ships based in New England killed and processed thousands of whales, bringing back thousands of barrels of their oil. Oil from the sperm whale was considered the best for lamps and spermaceti candles (a wax extracted from the head cavities of sperm whales) were found in every home. But hunting the whales was expensive, difficult, and dangerous, so cheap substitutes for whale oil were always sought. The first of these substitutes was an oil distilled from coal. In 1854, the Canadian geologist Abraham Gesner took out a patent on this fuel, calling it kerosene. Around this same time, a petroleum distillate, or unrefined natural mineral oil, was being sold in drug stores and tent shows under the name "Seneca oil" or "snake oil"— a foul-smelling medicine alleged to cure every ailment from cancer to cramps. The composition of snake oil formulations varied widely, but it frequently also contained other substances like camphor, red pepper, and turpentine. Given the fact that these cure-all products were basically hoax medicine, the term "snake oil" became synonymous with quackery.

Refining is the manufacture of petroleum products from crude oil. Crude is unrefined liquid petroleum. It is composed of thousands of different hydrocarbon compounds, all with different boiling points. As crude petroleum is heated, it changes into a gas. The gases pass into the bottom of a distillation chamber and become cooler as they move up the column. As the gases cool, they condense into liquids. The liquids are drawn off the distilling unit at specific heights. Heavy residual oil is collected at the bottom, raw diesel midway, and raw gasoline at the top. These raw oils are then processed further to produce the finished products. Crude oil can be processed and refined into petroleum, naptha, gasoline, diesel, asphalt base heating oil, kerosene, and liquefied petroleum gas.

At first, people extracted petroleum as a byproduct of salt wells, but George H. Bissell and Jonathan G. Eveleth of New York came up with the idea of drilling directly for oil. Colonel E.L. Drake of Titusville, PA invented the modern method of driving iron pipes, one length after another, down into a hole, thus keeping out water, quicksand, and clay and making it possible to extract the oil in a relatively uncontaminated form.

LYMAN STEWART (1840 - 1923)

Lyman Stewart was one of the U.S.'s first petroleum pioneers. His ability to find oil was so great that some people claimed he had a "nose for oil." In the heyday of California's oil strikes, Stewart had the reputation of being able to sniff the air and know if oil was available underground.

Stewart's father had a tannery in the village of Cherrytree, PA, near Titusville and he had wanted his son to follow in his footsteps. Stewart, however, had his own dreams. "Living within a few miles of (the oil fields of) Titusville," he explained, " it was natural that I should become interested in the new industry which was causing such excitement. My boyish enthusiasm became so strong, I couldn't resist the excitement."

The country's first oil well was drilled near Titusville, in 1859, and in the same year, Lyman Stewart invested his life savings of $125 in the purchase of an oil lease. He then lost the lease and all of his savings (when he and his partners could not secure the additional funding to actually drill). Stewart's professional career would often be a mixture of success and failure; many times he found himself on the brink of financial disaster, even losing his home at one point.

It was in the Titusville oil fields that Stewart met Wallace Hardison, and in 1877, the two men formed a partnership to buy up oil that would, years later, make them very rich men. Stewart and Hardison's partnership began with only a handshake. They built up modest oil fortunes, then sold out and moved to California in 1883 for greater opportunity. Stewart arrived in California at age 43 having already made and lost his first million.

Lyman Stewart's foremost interests were oil, his family, and the Presbyterian Church. A small, slender man of great dignity, he was not a stereotypical oil man. Always dressed immaculately, with a carefully trimmed beard, he

Figure 1: Portrait of Lyman Stewart

was soft-spoken and courteous. But behind his quiet demeanor was a shrewd businessman who constantly sought new oil leases and experimented with new products for new markets. He considered the refinery to be the key to the oil business.

On October 17, 1890, Lyman Stewart, Thomas Bard, and Wallace Hardison founded the Union Oil Company of California. The oil industry was barely thirty years old. The new company was headquartered in Santa Paula, the heart of the state's oil country. Though the partners were aware of oil's potential as an industrial and transportation fuel, no one could have foreseen the impact the automobile would have on oil demand or how oil would revolutionize train transportation and the nation's shipping infrastructure. Stewart, Bard, and Hardison worked to find or devise new uses for fuel oil because, as hard as it is to comprehend today, the Union Oil Company was often oil-rich and cash-poor.

Stewart was an oil hunter, a land buyer, and a petroleum producer who saw the advantages of having one integrated company produce, refine, and distribute oil. His research laboratory tried to find new ways to extract clear, nonsmoking kerosene from the smelly California crude. He built an oil pipeline that enabled his company to dramatically increase its oil output and bring oil more easily to the rest of the world. By the turn of the century, Lyman Stewart was the only one of the original three partners to remain in the business. The first gasoline powered automobiles appeared in the western United States in the early 1890s. By 1913, there were nearly 123,000 automobiles in California, prompting Union Oil to open its first service station (" gas stand" as it was called then) on the corner of 6th and Mateo Streets in Los Angeles. By 1925 Union Oil had more than 400 service stations on the West coast. Gasoline was still a sideline, though. Union's number one product was fuel oil.

Stewart served as the president of the Union Oil Company of California until 1914, when he was succeeded by his son, William. He continued to play an active role as chairman of the company until his death at age 83.

His obituary in *Petroleum World* stated, "His name was known wherever oil was spoken." Stewart was recognized as the father of oil in the Pacific coast region and given the title of "dean of western oil men". (Figures 1 and 2)

Figure 2: Hardison & Stewart Oil Company Stock Certificate

OIL WELL PACKER, PATENT NO. 230,080
Lyman Stewart, Titusville, Pennsylvania, July 13, 1880

An oil well packer that controls the leakage and flow of oil.

The packer's flange is secured to the tubing, a second flange is attached to the outside of the upper end of the tubing, and a telescopic cone wedge fits over the upper end of the tubing. A reducer is screwed upon the upper end of the cone section to receive the upper portion of the tubing.

HYDRAULIC MINING APPARATUS, PATENT NO. 209,661
George W. Cranston, San Francisco, California, November 5, 1878

This gold-mining device raises promising earth, sand and gravel by water pressure, and separates the gold particles from the earthy matter.

ROCK DRILLING MACHINE, PATENT NO. 130,412
John Cody, New York, New York, August 13, 1872

An improved steam drill that can operate in any desired position. With its grooved flanges, feed chain, piston rod, and swiveled screw wheel, the drill can be fed forward and then automatically rotates as it moves back after making a stroke.

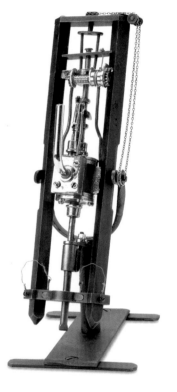

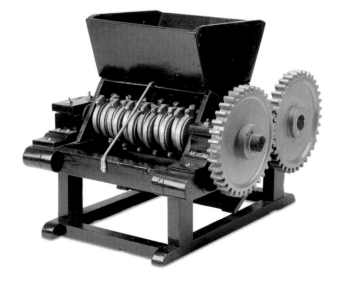

COAL BREAKER, PATENT NO. 219,773
Philip Henry Sharp, Wilkes Barre, Pennsylvania, September 16, 1879

A coal breaker that improves on the old design of two parallel rollers, studded with teeth, which crushed the coal. This breaker uses the splitting action of a series of reciprocating picks.

MACHINE FOR WASHING AND CONCENTRATING ORES, PATENT NO. 228,125
Frederick J. Seymour, Wolcottville, Connecticut, May 25, 1880

A machine that "embraces improvements and advantages for washing or concentrating auriferous and other ores." The rotating pan has a hollow upright shaft and central overflow. As the pan spins, the pan's agitators skim the surface of the water, working against the centrifugal force developed in the material and causing lighter matter to overflow the pan. The heavier materials, such as gold particles, settle near the rim of the pan where they are easily collected.

INDICATORS FOR MINING SHAFTS, PATENT NO. 193,462
Calvin O. Richardson, San Francisco, California, July 24, 1877

This invention warns miners to stand away when the cage is descending. The bell's clapper is supported by a spring shank, which lets the clapper vibrate when the bell moves. When it is attached to the hoisting rope of a mining shaft, the bell sounds as the cage descends.

APPARATUS FOR CRUSHING AND PULVERIZING ORES, PATENT NO. 262,652
Miles B. Dodge, San Francisco, California, August 15, 1882

In certain portions of the gravel mining regions of California deposits of gold-bearing gravels are found, in which the conglomerations are so hard that ordinary hydraulic mining operation will not suffice. Drifting out and then crushing breaks up the lumps and masses. Dodge's efficient design for a cylindrical rotary crushing mill has a series of interior solid crushing faces that are tangential to the circle of the carrying disks.

FURNACES FOR ROASTING ORES, PATENT NO. 207,890
David J. O'Harra, Reno, Nevada, September 10, 1875

A roasting furnace for desulphurizing and chloridizing ores. The furnace has two hearths, one above the other, with a connecting passage between them. The upper hearth desulphurizes the ore, and the lower one chloridizes it. Conveyers pass through both compartments, stirring up the ore and gradually conveying it through the furnace.

ORE STAMP MILL, PATENT NO. 232,878
Richard Bridewell, San Francisco, California, October 5, 1880

A stamp mill breaks up ore by pounding rather than grinding, for further processing or for the extraction of metallic ores. Bridewell's stamp battery minimizes both the power needed and the cost of construction.

CHAPTER 7

Toys

"The greatest invention in the world is the mind of a child."

—Thomas A. Edison

Prehistoric toys have been discovered on archaeological digs and found buried in ancient tombs. Wooden toys date back to the ancient Greeks, Romans, and Egyptians. Ancient Egyptians made balls from leather and dried papyrus reeds. Marbles were carved from black and white stones. Dolls were made from leather and papyrus, wood, bone, or ivory. Some dolls had jointed limbs and moveable jaws. Romans enjoyed board games and marbles, and created dolls and figures out of wood, wax, and clay. Children in early Greece played with wheeled wooden horses pulled by strings, clay human dolls and animal figures, and even primitive yo-yos. Kites have been flown in China for many centuries.

In the United States, class played a role in what toys children had access to. Poorer children's toys were frequently simple and handmade. Richer families were able to afford greater variety and better quality playthings for their children. Many toys were imported from England.

In the late 1600s, John Locke, a British philosopher, popularized the idea that toys helped in the learning process. As a result, by the end of the 1700s, there were educational toys for many different subject areas. The higher the social status, the more toys, free time, and formal education a child had. Poor children had access to few toys and often had to make their own. Middle class boys played with marbles, toy soldiers, and toy trains, while girls played with wood or porcelain dolls, doll houses, and jump rope. Some of the toy trains even had working engines powered by methylated spirits. In well-off Victorian families, children sometimes also had clockwork toys and rocking horses.

The Industrial Revolution changed the way in which toys were made and marketed. Mass produced toys were cheaper to make and consequently more affordable to purchase. Following the Civil War, many of the factories used to make weapons and military supplies were turned over to toy production, which is why this period is known for its distinctive tinplate and cast-iron toys. The mechanical methods of stamping tinplate, molding papier mâché, and casting iron revolutionized the toy industry, with toy makers still often adding hand-painted details.

The American clockwork toy was most popular between 1860-1880. Many patents were issued for dancing figures, and clockworks added "life" to lively dancers. Clockwork toys were wound with a crank or a key. The external parts of these toys were generally made from painted tin, wood, cast iron, or cloth, but the internal workings were made first from brass, and after 1890, from steel.

Clockwork toys characterized various aspects of life during this time period. Dancing and walking dolls, merry-go-rounds, velocipedes (what we now call a bicycle), locomotives, and even riverboats with paddle wheels captured the flavor of day-to-day life of the time.

Tin toy manufacturing in the United States goes back to 1850 when there were at least fifty toy manufacturers in business. Clockwork toys were a byproduct of the clock industry, with production centered in Connecticut. The state was even known as the "toy center of New England."

ENOCH RICE MORRISON (1812 - 1889)

On August 6, 1861, Enoch Rice Morrison of New York City was granted Patent # 33,019 for a self-propelled locomotive apparatus. His walking doll was called an "Autoperipatetikos" from the Greek words, auto, meaning self-propelled and *peripatetikos*, meaning to walk about. The moveable legs and feet supported a clockwork mechanism, which provided a walking action. The mechanism was concealed within a cardboard underskirt, above the wooden base from which the legs protruded. The doll wore a full-length, hoop-skirted dress. The height of the doll was approximately ten inches. Originally, the doll came with an imported china or porcelain head, but later, a fabric-based head was used. The toy was manufactured by Martin and Runyan of New York City.

The "Autoperipatetikos" was also patented in England, in 1862, but there the doll had a slender woman's body.

A rare American version of this doll is the "Walking Zouave" named after the Zouave Militia. In 1859, Elmer E. Ellsworth organized and trained, in Chicago, a volunteer militia of Zouaves, patterned after the original French infantry regiments of the same name organized in Algeria in 1831. At the beginning of the Civil War, Ellsworth recruited a

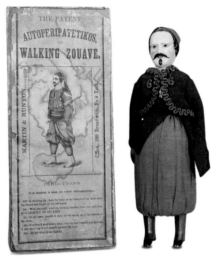

Figure 1: Autoperipatektikos, Walking Zouave

regiment of New York City firemen which mustered into Union service in 1861. Eventually many other Zouave regiments were organized. The uniform of the Zouave regimen consisted of a headdress or red fez with a gold tassel and a blue jacket with gold trim. A long blue sash around the waist held a knife or similar weapon, and baggy scarlet trousers were tucked into yellow or white gaiters. (Figure 1)

LOCOMOTIVE APPARATUS, PATENT NO. 33,019
E.R. Morrison, South Bergen, New Jersey, August 6, 1861

Morrison's Locomotive Apparatus consists in a frame or box supported upon legs and feet, which by the action of a spring or a clockwork mechanism creates an alternating step-by-step movement resembling walking. This invention is intended to be applied to dolls and toys, but may be applied for other uses.

The clockwork mechanism actuates the movement of the legs. The apparatus maintains its balance because each foot has a bar that extends across to the other side. It walks with a peculiar and fascinating lurching movement.

LUDWIG GREINER (1796 - 1882)

The mass production of papier mâché dolls began in the early 19th century and continued into the early 20th. China invented papier mâché, and its use goes back as far as the Han Dynasty (202 BC - 220 AD). The term papier mâché is French and can be literally translated as "chewed paper." It was commonly a mixture of paper, sawdust, plaster, and glue, typically pressed into a mold for shaping. The molds for dolls allowed them to be mass-produced and made affordable to many families. Germany, France, and the United States were among the leaders in the production of papier mâché dolls.

Figure 2: Original Greiner Antique Doll

In order to produce a more lifelike appearance, some papier mâché doll heads had a thin overlay of tinted wax.

The first American-made papier mâché dolls were made by Ludwig (sometimes known as Lewis) Greiner. Born in Germany, Greiner immigrated to the United States in the 1830s and moved to Philadelphia. He produced papier mâché dolls from 1840 to 1874. He was listed in the 1840 Philadelphia city directory as "toy man." Greiner was issued the first dollmaking patent in the United States, for his method of molding papier mâché heads. His early dolls are labeled with a paper label that reads "GREINER'S IMPROVED PATENT HEADS Pat. March 30th '58." Greiner dolls ranged in size from 13 to 36 inches tall. He painted his doll heads in oil and finished them with varnish. The earliest dolls were all brunettes. Ludwig Greiner's doll factory in Philadelphia produced only the doll heads and left the buyer to make the body and clothing at home. Greiner dolls became among the most popular and recognizable papier mâché dolls ever produced. (Figure 2)

CONSTRUCTION OF DOLL HEADS, PATENT NO. 19,770
Ludwig Greiner, Philadelphia, Pennsylvania, March 30, 1858

The Greiner invention consists in applying linen, silk, or muslin using his patented process to make doll heads, that when dropped the heads would not break.

The manner of preparation is as follows: One pound of white paper, when cooked, is beat fine, the water is pressed out, so as to leave it moist. To this is added one pound of dry Spanish whiting, one pound of rye flour, and one ounce of glue. This is worked until it is well mixed. It is then rolled out to the required thinness and cut into narrow strips required for use in the molding process.The combination of using either linen or muslin with the compounded paste material is then molded into a doll's head using the Greiner mold and his patented process. Oil paints can then be used to paint the completed doll's head.

This modern day replica of the doll's head was created and painted by artist, Rhea Fletcher of Salem, SC, by using the original Greiner mold and his patented process for creating doll heads.

JESSE ARMOUR CRANDALL (1834 - 1920)

Jesse Crandall was an American inventor and toy maker. He took out over 150 patents on toys in his seventy-five years of inventing. His father, Benjamin Potter Crandall, was also a toy maker, and so were three of Jesse's brothers. Unlike his brothers, who remained primarily associated with their father's toy business in New York City, Jesse left the family business and soon after the Civil War started his own company in Brooklyn. He was called "The Child's Benefactor" and used this as his trademark and slogan. His specialties were hobby horses, rocking horses, velocipedes, and board games.

Figure 3: Original Shoo-Fly style Antique Rocking Horse

Crandall's father began selling baby carriages in the 1830s, billed as "the first baby carriages manufactured in America." Jesse designed a tool to drill the ten evenly-spaced holes in carriage wheels when he was only eleven years old. He was issued a number of patents for improvements and additions to the standard models of baby carriages. These included a carriage with brakes, a folding carriage, a carriage with an oscillating axle, and designs for carriage parasols and an umbrella hanger.

Crandall is credited with inventing the Shoo-fly design of the rocking horse in 1859. The Shoo-fly used two pieces of wood in the shape of a horse attached to the sides of a box or seat, mounted on bow rockers. It had the advantage of being easier for young children to balance upon. In the 1860s, Jesse and Charles Crandall patented an interlocking system using tongue and groove blocks to produce acrobats, circus characters, and buildings. In 1861, he was issued a patent for a spring-loaded rocking horse and through the years was also issued a number of patents for rocking toys, alphabet blocks, and construction toys. (Figure 3)

SKIPPING APPARATUS, PATENT NO. 169,625
Jesse A. Crandall, Brooklyn, New York, November 9, 1875

A substitute for jumping or skipping ropes. The belt is placed around the child's waist, and when the child turns the cranks, the hoops revolve around the body for the child to jump over.

"The apparatus, when properly handled, will afford much amusement to children, and give occasion for graceful motions."

MACHINE FOR MAKING TOY TORPEDOES, PATENT NO. 165,960
Thaddeus H. Spear, Gardiner, Maine, July 27, 1875

This invention automated the production of toy torpedoes. A toy torpedo is a little sack of paper twisted together at the top end, with a small amount of fulminating powder and a few grains of gravel inside it. Throwing the sack on the ground brings the gravel and the fulminate together and causes the desired explosion.

Toy torpedoes, or impact-actuated exploding fireworks, were very popular and fireworks manufacturers produced many sizes and kinds, with a variety of explosive compositions, before they were finally banned in the late 1950s. Today's "Snap n Pops" are mere shadows of the toy torpedoes once available.

WRESTLING TOY, PATENT NO. 62,455
James T. Walker, Palmyra, New York, February 26, 1867

Walker named this toy the "Lilliputian Wrestlers." Pressing the lever throws the figures down, "causing them to perform various evolutions. The many movements greatly resemble the wrestling of two people with each other, whereby much amusement can be afforded to children and others."

MECHANICAL TOY WALTZING FIGURES, PATENT NO. 128,164
Edward E. Newell, Bristol, Connecticut, June 18, 1872

These beautifully hand-carved figures are just part of a mechanical toy that could traverse a floor while spinning. "Upon the winding up of the mechanical move-ment, the dancers will gracefully waltz around over a floor or similar plane for an extended period of time. The braking system can be adjusted to vary the speed of travel depending upon whether the dancers are on a bare or carpeted floor."

The toy uses a vertical and revolving shaft, with a mechanical movement. The carriage has a propelling spring and a braking mechanism.

This invention may be adapted to any mechanical toy carriage with equally good results.

SWING, PATENT NO. 109,165
Lucius Winston, Pontiac, Illinois, November 8, 1870

Winston invented a new type of seat for either a single or double swing, securing the seat so it stays horizontal, no matter how high it rises in the air. Its path is on an inclined plane, and not on an arc of a circle, as are other swings.

TOY WAGON, PATENT NO. 125,126
Henry W. Eastman, Baltimore, Maryland, April 2, 1872

A toy wagon that folds into a small package to save space.

The two sides of the wagon are hinged to the bottom of the wagon box with metal hinges and screws; removing the screws lets the sides fold down. The head board and tail board are also easily removable. Metal trunnions and bolts are used in the construction of the wagon box and securely attach the wheels.

TOY BOAT, PATENT NO. 133,250
Joseph W. Pilkington, Bridgeport, Connecticut, November 19, 1872

This toy boat has an oar at the stern of the boat with a vibrating clockwork figure, which gives a sculling motion to the oar and propels the boat. The oar is fitted with a sleeve so the boat can be steered to the left or right.

METAL TOYS, PATENT NO. 178,366
William A. Harwood, Brooklyn, New York, June 6, 1876

A "contrivance of the support by which the horse is mounted on the wheels, so the horse will be elevated considerably higher than the common method . . . By mounting it higher, the horse looks larger, and makes a more attractive toy."

Using a narrow strip or strips of metal, or one or two pieces of wire, coiled, arched, corrugated or otherwise shaped or bent, makes a lighter support, and, at the same time, stiffens the metal.

Harwood's toy horse was put into production by the Merriam Manufacturing Co. in Durham, CT. Bernard Barenholtz in his book, "American Antique Toys," called this toy "the most graceful of all American tin toys."

TOY FISH, PATENT NO. 84,628
Robert Hunter, New York, New York, December 1, 1868

Robert Hunter of 9 Brevoort Place, in the city of New York, a doctor of medicine, invented a Propeller for Toys, intended more particularly for mechanical fish, but also applicable for toy boats.

The peculiarity of this invention consists in a mechanical fish with a tail made of thin elastic material, which vibrates by means of a coil spring and ordinary gearing. Winding the crank moves the tail from side to side. "This gives the fish the appearance of life when placed in water, and causes it to swim with precisely the same motions of a live fish."

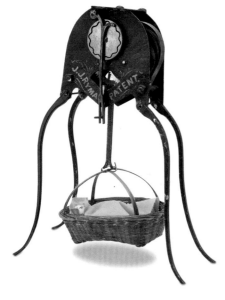

AUTOMATIC SWING, PATENT NO. 79,147
John J. Rymal, Rochester, Minnesota, June 23, 1868

A swing for babies or children, with a noiseless vibrating motion. The swing can be placed on a floor, hung from the ceiling or hung on brackets from a wall. The soothing motion of the rocking back and forth of the cradle is accomplished by use of a clock work mechanism that can be wound up to power the swing for an extended period of time.

David Saint, a designer working for the Graco Metal Company, has been credited with introducing the first automated baby swing in 1955. It appears that the real inventor, Rymal, was ahead of his time by more than 80 years. Today, automated baby swings are powered either by batteries or by electricity.

AUTOMATIC TOY, PATENT NO. 140,883
Henry L. Brower, New York, New York, July 15, 1873

The "Acrobat" is a toy image of an athlete with a bar, combined with a wind-up clockwork mechanism. Special gearings give the bar motions that induces a variety of fantastic movements. The figure performs partial or complete revolutions alternately in opposite directions around the bar, and the movements vary irregularly.

AUTOMATIC TOY DANCERS, PATENT NO. 143,121
Henry L. Brower, Yonkers, New York, September 23, 1873

The dancer is mounted on a lever and operated by clockwork. A spring on the lever balances or partly balances the weight of the figure. "The joints allow great freedom of motion, and the elasticity of the lever gives the figure an active jumping motion, with a descent a little more rapid than is due to gravity alone." Sadly, these minstrel show/black face figures were all too common and acceptable at the time.

TOY VELOCIPEDE, PATENT NO. 96,972
Ernst Santin, New York, New York, November 16, 1869

A new toy velocipede which is propelled by spring power, and gives motion to a jointed figure of a man or other figure. As the wheels of the velocipede turn, the extremities of the figure move. Thereby a very amusing toy is produced, especially when the figures of animals, such as monkeys, dogs, etc. are placed upon the vehicle, so that their motion will correspond with those of the animal they represent.

TOY CARRIAGES, PATENT NO. 86,814
John Condell, Archelaus Condell, Plainville, Connecticut,
February 9, 1869

A toy carriage which moves the toy figures as the carriage's wheel is trundled, which "causes the arm to reciprocate up and down upon the post causing the images to rise, advance heads towards each other, and back in an entertaining motion, with each revolution of the wheel."

TRUNDLE TOY, PATENT NO. 174,265
Elam B. Long, Christopher Eisenhardt,
Philadelphia, Pennsylvania, February, 29, 1876

A toy with a driving wheel and handle, which the child pushes over the floor. Dolls, toy animals or other objects may be placed on the platform. To increase the attractiveness of the toy, a sound like that of a watchman's rattle will be given when the toy is being used.

CHAPTER 8

Home

"A house is not a home unless it contains food and fire for the mind as well as the body."

—Benjamin Franklin

The Industrial Revolution saw the U.S. transformed from a predominantly agricultural nation into a more industrialized one. During this period, the population also shifted from being primarily rural to more people moving into urban settings.

After the Civil War, the population of U.S. cities steadily increased. In 1860, about fifteen million people lived in cities, but by 1900, that number had doubled to 30 million. Two factors that influenced this trend were the arrival of immigrants to the U.S., and farmers giving up farming and moving into cities to find work. Cities grew as factories grew. People looking for decent-paying jobs were forced to move to cities where most of the factories (and jobs) were located. This new city life could be difficult and dangerous. Workers found it hard to afford decent housing because of their low wages, and many were forced to live in overcrowded, run-down buildings called tenements. Most tenements had no windows, heating, or even plumbing. Fire was a constant threat because the wooden buildings were built so close together. Poor people lived downtown in the oldest, most run-down sections of cities. Farther out, the middle class (doctors, lawyers, office workers, and skilled craftspeople) lived in row houses or newer apartment buildings. The rich built mansions on the cities' outskirts—large houses with big lawns and trees.

Both the Civil War and the Industrial Revolution also impacted the role of women. Prior to the Industrial Revolution, a married woman's role was as a housewife and mother, and for single women, only jobs like teachers, nannies, and housekeepers were available. Housework required enormous amounts of physical effort. Women ran the house, raised the children, made the family's clothing, cared for sick family members, and prepared (and often grew) much of what the family ate. The Industrial Revolution made it acceptable for married women to work outside the home and to be more independent. This process accelerated with the Civil War. Hospitals, in particular, needed women's help. Sixty percent of women who worked outside the home worked as domestic servants in the 1870s. By 1900, the number of women engaged in domestic services had declined to one-third while an increase was seen in factory, office, retail, teaching and other professional jobs. The massive migration from farm to city, begun as industrialization rapidly expanded in late 19th century, may be the most important change in women's lives and social and work roles during this period.

The nineteenth century was a period in which the American population began its migration from producers to consumers. The rapid growth of communications, transportation, and mass production had somewhat of an equalizing effect on rural and urban populations. Consumer goods became available by mail order and in large department stores, bringing some of the products and conveniences of urbanity to rural communities. By the late 1800s, new inventions began to appear to assist women with manual and domestic tasks, the sewing machine being one important example. Patented in the 1850s, the sewing machine was initially so expensive only factories that mass-produced clothing could afford them. Manufactured clothing became widely available as a result of the Civil War, where army uniforms had been needed. After the war, men's clothing continued to be mass produced in this way, while women's garments remained homemade and by independent seamstresses until the early 20th century.

In the 1800s, before the end of the Civil War, African-American slaves were prohibited from filing patents. Former slaves and free African-Americans were, of course, allowed to receive letters of patent just like anyone else. After the

Emancipation, and the end of the war, increasing numbers of African-Americans began filing for patents. Perhaps best-known of these inventors was George Washington Carver, a former slave, who came up with alternative uses for peanuts and sweet potatoes. Another inventor of the period was Jonas Cooper. A janitor living in Washington, DC, in May 1883, Cooper received a patent for lightweight, removable window shutters.

SHUTTER FOR WINDOWS, PATENT NO. 276,563
Jonas Cooper, Washington, District of Columbia, MAY 1, 1883

Previous removable window shutters were cumbersome and heavy, costly and difficult to repair. These new shutters ran smoothly in fixed grooves, two or more to a window, and could be quickly removed or inserted. They were light, firm and cheaply constructed. Jonas Cooper was a rare example in the Rothschild Patent Model Collection of an identified African-American inventor. He was one of a very small group of African-American inventors that received a patent during this period of time.

MELVILLE R. BISSELL (1843-1889)

In 1876, Melville R. Bissell of Grand Rapids, MI invented the Bissell Carpet Sweeper. (Figure 1) Bissell and his wife, Anna were partners in a crockery store. As the story goes, most of their merchandise was very fragile glass and china products which would arrive in crates packed with sawdust. This packing material would frequently spill onto the floor of their shop and needed to be constantly cleaned up. In sweeping up the wood dust, Melville Bissell used a carpet sweeper that employed wheels and rotating brushes which swept the dirt out of the pile in rugs. Bissell was frustrated with the sweeper and started thinking about a better design. The Bissell device he came up with also used floor wheels to drive a brush, but employed a much improved reduction gear. The bristles also bent slightly as they brushed the carpet. As they rotated off the floor, they lifted whatever debris was in their path and deposited it into a compartment.

Figure 1: Portrait of Melville R. Bissell

The dirt was then emptied by simply opening the container and shaking it into the trash.

Melville Bissell died of pneumonia in 1889 at the age of 45. Upon his death, his wife Anna became head of the company, becoming the first female CEO in America. She defended and championed the company's patents and marketed Bissell carpet sweepers successfully throughout the U.S., Canada, and Europe. By 1899, she had created the largest corporation of its kind in the world. Anna Bissell introduced progressive labor relations policies, including workmen's compensation insurance, and pension plans to her employees, long before these practices were widespread. In the late 1890s Queen Victoria was known to insist that her palace be cleaned by BISSELL carpet sweepers every week, a practice she called "Bisselling" her floors. The Bissell Co. is still in existence today and is still family owned. (Figure 2)

CARPET SWEEPER, PATENT NO. 335,010
Melvin B. Bissell, Walter J. Drew,
Grand Rapids, Michigan, January 26, 1886,
Assignors to the Bissell Carpet Sweeper Company

A class of carpet sweepers in which the drive wheels (the ones that turn the rotary brush shaft by frictional contact) are located outside the end walls of the sweeper casing, the latter being also provided with an encircling elastic band for preventing the drive wheels and said casing from coming in contact and marring furniture or wall surfaces.

Over 125 years after this patent was assigned, the Bissell Company is still located in Grand Rapids, Michigan, manufacturing carpet cleaners and numerous other cleaning products that originated with the early patents of Bissell and Drew in the late 1800s.

Figure 2: Bissell Carpet Sweeper Co. Factory

FOLDING FURNITURE

After the Civil War, "folding" furniture became increasingly popular in the U.S. This portable furniture provided a practical solution to the problem of crowded living. Increased urbanization and cramped living quarters called for extra ingenuity and space-saving.

The first rocking chairs were introduced beginning around 1800. The curved slat fastened to the feet of the chair enabled it to be rocked back and forth. Between 1831 and 1905, the U.S. Patent Office issued more than 300 patents specifically for rocking chairs. The folding rocking chair folded flat for ease of shipping, storing, and moving from room to room

or even outdoors. Between 1855, when the first U.S. Patent for a folding chair was issued, and by the end of the century, more than 355 patents were granted for folding chairs. Mass produced and modestly priced, the folding chair was affordable for almost every pocketbook and could also be finished to appeal to a wide variety of tastes.

In 1876, the U.S. International Exhibition was held in Philadelphia to celebrate the one hundredth birthday of this country. This event became known as "The Centennial." At the Centennial, the furniture displays were among the most elaborate and popular. The American furniture shown at the

Exhibition was generally more functional than aesthetic and was noteworthy for its high degree of specialization. Quite frequently, a company carried only one kind of furniture—just chairs or desks, for example.

One such company was the Vaill Chair Co., owned by Edward W. Vaill (E.W.) of Worcester, MA. They specialized in the manufacturing of folding chairs. The Vaill Co. had a display of its folding chairs at the 1876 Centennial Exhibition.

Many patents for folding chairs and folding rocking chairs were assigned to E.W. Vaill. One patent, for the Improvement in Folding Rocking Chairs, was patented by Ephraim Tucker, but assigned to Edward W. Vaill. Vaill himself held at least five patents for folding chairs. Minimally, at least another five patents for such chairs were submitted by other men, presumably employees in Vaill's factory, but assigned to Vaill.

Edward W. Vaill manufactured chairs in his factory in Worcester, MA from 1861 until 1891. The Union army had a great need for camp furniture and Vaill's company manufactured a folding camp chair that was easily transportable. As a result, the military commissioned many folding chairs for their use. In 1863, Vaill devoted his entire company to the manufacture of these chairs. With the end of the war, the folding chair evolved into a form more suited to general use and it became a popular furniture staple for the drawing room, library, veranda, church, public hall, seaside, and shipboard. Vaill chairs often had needlepoint tapestry or carpet and fringe decorating the arms or front of the seat. The Vaill Chair Co. is believed to have been the largest folding chair company of its time. (Figure 3)

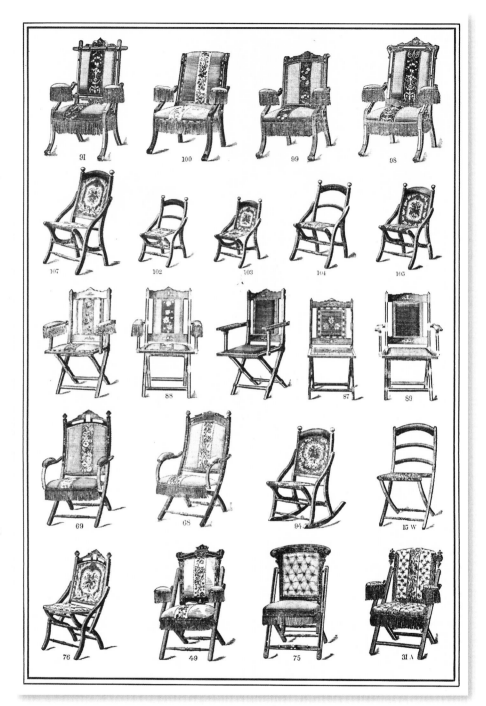

Figure 3: Advertisement Vaill Chair Co.

FOLDING ROCKING CHAIR, PATENT NO. 205,015

Ephraim Tucker, Worcester, Massachusetts, June 18, 1878, Assignor to Edward W. Vaill

This ingenious folding rocking chair uses pivots and hinges to reduce the size of the folded chair to a minimum. so it is compact and takes up little space in an overcrowded room.

COMBINATION FURNITURE

Big cities in the mid-1800s, where space was at a premium, created the need for innovative, multi-use inventions. Built-in, folding, extension, and multipurpose furniture was designed with an eye toward utilizing existing space to its best advantage. Many of these inventions were created in response to the loss of living space as more people moved into smaller urban homes and apartments. The Patent and Trademark Office even designated a specific category for "combination furniture."

Many patented inventions of this period included beds as part of their concept. Numerous examples of combination sofa-beds, table-beds, wardrobe-beds, and bureau-beds can be seen in patent records. The more ingenious ideas included Stephen Hedges' combined table and chair (Patent # 10,740) and Charles Hess' patent for a combined piano, couch, and bureau (Patent # 56,413), Hess envisioned his design being

RAWSON'S IMPROVED
Combined Chair and Step Ladder.
AN ARTICLE OF GREAT UTILITY AND UNIVERSALLY LIKED.

Made of Oak, in the most substantial manner, and sold at the price of an ordinary Step Ladder.

Always ready, and always in order.

Made in the form shown in the cut, for household use, and also upholstered in a variety of styles for library and office use, either with or without arms.

Exclusive Territory given to Responsible Dealers and Experienced Salesmen. Address
C. B. RAWSON, - 89 Exchange Street,
WORCESTER, MASS.

Figure 4: Advertisement C.B. Rawson

used in an apartment with a parlor needed for activities during the day and for sleeping at night. George V. Leicester received his patent for a combination toilet-bedstead (Patent #96,599), while John McDonald invented a combined bed and musical-instrument board (Patent #97,101). McDonald explained: "My invention has for its object to furnish a keyboard musical instrument, which shall be so constructed, that it may be opened up to serve as a bed, and which, when closed, shall have every appearance of, and may in fact be a real instrument, suitable to be placed in a parlor or sitting room."

Charles B. Rawson received at least two patents for his invention for a combined chair and stepladder (Patent # 133,594 and Patent # 165,688), a patent for a jointed curtain pole for bay windows (Patent # 269,462) and for Making Match-Splints (Patent # 298,502). (Figure 4)

COMBINED CHAIR AND STEP LADDER, PATENT NO. 165,688

Charles B. Rawson,
Worcester, Massachusetts,
July 20, 1875

Rawson's invention is a very fine example of combining two separate objects into a single piece of furniture to save space. With the seat platform down, the chair resembles an ordinary household chair. However, raising the seat platform turns the chair into a functional working ladder, as the step rungs that connect the legs together combine with steps built into the underside of the platform seat to become a ladder "constructed to ensure complete safety to the user."

CANE SEAT, PATENT NO. 80,300

George W. Martin, Boston, Massachusetts, July 28, 1868

Typically, a cane chair seat is only caned on the top side of the seat. After much wear and tear, one cane strip typically breaks or wears out, and then the adjacent ones do, rendering the piece unusable.

Martin cleverly remedied this problem by caning the seat exactly in the same way on both the top and the bottom portions of the seat. Once the seat's cane became unusable, you simply reversed the seat instead of repairing or recaning it.

BOOK CASE, PATENT NO. 182,669

William Homes, Boston, Massachusetts, September 26, 1876

This portable book case is put together without bolts, screws, or nails. Instead, it uses springs, knobs, handles, hooks, and a grooved back board which can be firmly locked, but can be easily taken apart with ease to be moved.

REVERSIBLE LOUNGE, PATENT NO. 62,349
Samuel Lloyd, Washington, District of Colombia, February, 26, 1867

The Reversible Lounge is simple in construction but yet very clever in design. The head of the lounge can be reversed either to the left or right side, depending on which side is desirable, and allows for "equal wear on the resting portion of the lounge." No tools or implements are needed to make the change.

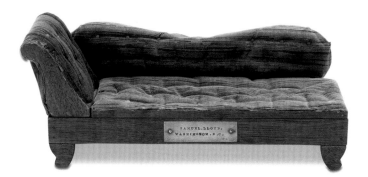 

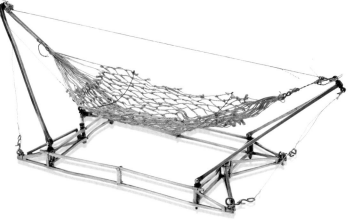

HAMMOCK FRAME, PATENT NO. 182,049
Juan B. Aric, Brooklyn, New York, September 12, 1876

An easily folding hammock frame, made of rigid metal side rails and hinged sectional end rails, with hinged braces for spreading the frame, folding derricks, and guy ropes for suspending the rope hammock.

SOFA BED, PATENT NO. 113,576
Charles C. Schmitt, New York, New York, April 11, 1871

An early sofa bed, with the length of sofa constituting the breadth of the bed. The sofa has a hollow or double back which conceals the mattress and also forms part of the bed bottom, and the bed headboard.

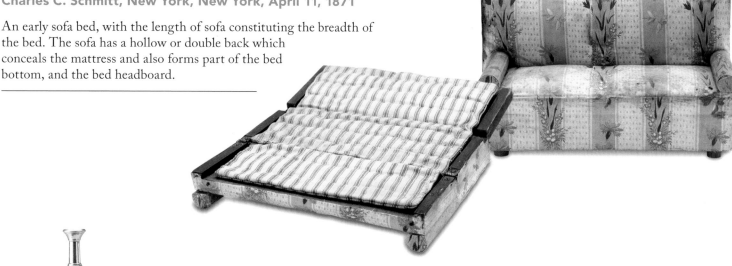

LAWN SPRINKLER, PATENT NO. 220,277
Frank N. Forster, Buffalo, New York, October 7, 1879

Forster's invention relates to that class of lawn sprinklers which consist of a vertical pipe supported by a base, and having a distributing nozzle at its upper end.

His design for a lawn sprinkler uses a flared nozzle and an inner plug with spiral water passages, to distribute the water in a fine spray.

The nozzle is simple and cheap to construct and not liable to become clogged by any solid impurities the water may carry.

AUTOMATIC FAN, PATENT NO. 179,211
Aaron Mattox, South Bloomingville, Ohio, June 27, 1876

A fly fan powered by clockwork that uses a flexible spiral connecting rod to join the shaft to the clockwork, spring stops, and folding arms. After winding, the four brushes produce a fanning motion which keeps flies away from a table, bed, or cradle.

This mechanism can also be used in the same manner to keep away mosquitoes.

PORCELAIN WATER CLOSET, PATENT NO. 215,495
Richard H. Watson, Philadelphia, Pennsylvania, May 20, 1879

Watson invented certain new and useful improvements in porcelain water closets with the object being to secure a greater purity and a more effective ventilation. His water closet is constructed out of a single piece of porcelain, with no joints. Note the internal curvature in the trunk of the closet, along with the inwardly shelving flange and pipe connections.

SPRING SHADE ROLLER, PATENT NO. 224,596
Daniel E. Kempster, Boston, Massachusetts, February 17, 1880

The conventional household window shade of today is based upon Kempster's invention of 1880.The object of his invention was to provide a spring curtain or shade roller which would lock automatically, without any manipulation whatever, by merely letting go of the shade either in its descent or ascent, and would not lock if the shade is permitted to ascend slowly under the influence of the spring. The combination of using a spiral winding spring, a cam disk fixed to the spindle along with a frictional locking devise provided for the successful operation of Kempster's invention.

SEWING MACHINE, PATENT NO. 19,723
James Sangster, Amos W. Sangster, Buffalo, New York, March 23, 1858

The loop stitch is extremely difficult to accomplish on a sewing machine. This invention addresses several issues related to the loop stitch, such as adjusting the feed motion to allow for variable stitch length, the way the loop is held while the needle passes through it, and the way the loop is protected from other parts of the machine.

REFRIGERATOR, PATENT NO. 62,860
Peter Lawson, Lowell, Massachusetts, March 12, 1867

To keep a refrigerator cool,it is important to introduce and remove articles from refrigerators quickly and conveniently. This revolving shelf, operated by gearing, allowed articles to be removed and replaced without having to rearrange other articles.

LAWN MOWER, PATENT NO. 215,366
Andrew Jusberg, Galva, Illinois, May 13, 1879

Jusberg's improvements to lawn mowers of 1879 may actually be one of the first patents for a Rotary Lawn Mower. The device is propelled over the grass by the two long wooden plow handles. As the driving wheel rotates, the gear wheel, which is connected to the cutter head plates, "creates a rotary motion to the blades or cutters that sever the blades of grass." Some of the advantages of the rotary mower over cylindrical mowers are the ease of cutting grass under low shrubs, bushes or trees, and cutting grass close up against sidewalks, posts, or other obstacles.

CHAPTER 9

Kitchen

"If you can't stand the heat, get out of the kitchen."

—Harry S. Truman

In the early 1800s, kitchens were places for preparing and cooking food, but also as a source of heat for the house. Before the invention of the cast-iron stove in the 1820s, cooking was done over an open hearth. Initially, only the rich could afford stoves, but by the 1850s, most homes (except those on the frontier and of the very poor) had a cast-iron stove. It was during the 19th century that many Americans began to buy their food for the first time, rather than grow it themselves. Urbanization and industrialization made food consumers out of formerly food producers. Modern transportation and food preservation techniques lengthened the seasons and introduced variety. In the late 1800s, food options for a family depended on whether you were rich or poor, urban or rural. Rural people continued to produce more of what they ate.

Before the 20th century, people ate an unvaried diet. Weather conditions, crop cycles, and poor transportation made only a few foods available at any time, though the foods changed with the seasons. Lower and middle class families ate bread, cheese, butter, porridges, eggs, raw fruits and vegetables in season, and preserved fruits and vegetables out of season. They drank beer, cider, milk, tea, or coffee. Water was often undrinkable.

Food arrived in kitchens unprepared. Shoppers returned from the market with live chickens to kill or dead ones to pluck. Fish had scales, hams had to be soaked or blanched, green coffee needed roasting and grinding, whole spices needed grinding and sifting. Flour, too, had to be sifted, as it might be full of impurities. These tasks began to disappear around the beginning of the 20th century as manufacturers took over the work of food preparation. Prepared food in cans and boxes, the result of mass production and mass distribution, entered the kitchen.

In cities, the houses of the wealthy had separate rooms to serve as kitchens, often staffed by servants, and separate rooms for dining. In the South, slaves or servants cooked. In all but the poorest Southern households, slaves did the bulk of the cooking until the Civil War. For the urban poor, living in tenements with families limited to one or two rooms, kitchens had to serve as workspaces, sleeping quarters, and living areas. With no sewer systems, filthy water overflowed into pump wells that provided water for cooking. Getting water for cooking and cleaning was a labor-intensive task. Most city households had to haul or pump water from a cistern or well until municipal water systems began to pipe water into city homes. New York City didn't receive piped water until 1842.

Dishwater was reused by washing the dirtiest and greasiest items last. China and dishes were done first, followed by milk pans, utensils, and finally the pots, roasters, and kettles. Fresh hot water was added at each load. Doing the dishes required hauling a great amount of water. The cold water had to be carried to the stove for heating and then carried to the dishpans. Finally, dirty water was carried outside to be dumped, often into the garden or the outhouse.

Thomas Moore patented the first domestic icebox in 1803, but many of the first iceboxes were homemade. Iceboxes gave homes a way to keep food cold. Fresh meat, dairy, and fruit could be kept in good condition much longer if kept cool. By the late 1800s, the numbers of homes with refrigerators and the amount of ice used had increased dramatically.

Thousands of patents were granted for kitchen utensils designed to ease the time and burden of food preparation, but it wasn't until mass production and a national distribution system that they were affordable for most families. Page after page of mail order catalogs were dedicated to tools like cherry pitters and apple peelers, enameled steel pots and kettles, meat choppers, and butter churns. Two kitchen innovations that were widely used were the the cast iron stove and the inexpensive, geared metal Dover eggbeater. The eggbeater greatly reduced the time and effort needed to beat, whip, or froth eggs. These devices made a significant impact on the way Americans cooked.

Before 1860, the fireplace was the heart of every room, and the kitchen fireplace with its open hearth was the center of the house. Many houses, especially in rural areas and small villages, heated only one room that served as kitchen, dining room, living room, workshop, and sometimes, bedroom. The kitchen fireplace dominated women's lives. Only constant fire-tending could keep a hot fire going, and wood had to be felled, chopped, and carried into the house. The job of cooking on those fires was hot and dangerous. Burning cinders flew from unscreened kitchen fires and in summer heat from the fire could be unbearable.

Cast-iron cooking stoves lessened much of the hazard and difficulty of fireplace cooking. The raised cooking surface meant less bending and heavy lifting than cooking on a hearth. Cookstoves were safer than hearth fires. House fires decreased as cookstove use grew. Through the middle 1800s, stove prices ranged from five to twenty-five dollars, at a time when the standard pay for common laborers was one dollar a day, and a barrel of flour, enough to last a family of five for eight weeks, cost between five and six dollars.

The cast-iron cooking stove could be considered the single most important domestic symbol of the 19th century.

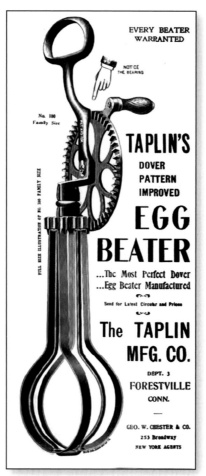

Figure 1: Advertisement for the Taplin Mfg. Co.

Benjamin Franklin's stove, invented during the 1740s, was intended for room heating, not cooking, and was not enclosed. The original Franklin stove had an open hearth, with channels through which the warm combustion gases could pass (so as to provide additional heat) before being vented out the chimney. See more of Franklin and his stove in Chapter 4: Heat, Light & Fire. Cooking and heating stoves had replaced the open hearth in most homes by the close of the Civil War. Stoves were economical, portable, and efficient. They required less fuel than fireplaces and were cheaper to install.

The Dover rotary eggbeater was the first mass produced eggbeater. Sears at the time claimed it to be the best beater made, and it set the standard for all others. When the first Dover eggbeater was patented, on May 31, 1870, one hundred and forty eggbeater patents had been granted. Many of these were manufactured, but just as many were never made.

Henry Marchand's Atmospheric and Vacuum Eggbeater did not resemble the Dover model. Marchand, in the spirit of invention, believed he had a better idea. Whether his invention was actually produced is something we don't know, but we do know that his design never became what we think of when we think of an eggbeater. Sales of hand-held eggbeaters peaked in the late 1940s and early 1950s. After that, the old style eggbeater gave way to the electric food mixer. The kitchen tools that had been new and unproven at the end of the 1800s had by the early 1900s become necessities. Kitchen tools modernized housekeeping, making it easier and lessening the time spent in the kitchen. (Figure 1)

ATMOSPHERIC AND VACUUM EGG BEATER, PATENT NO. 214,936
Henry E. Marchand, Pittsburg, Pennsylvania, April 29, 1879

This reciprocating tool beats eggs by using pressure to force them back and forth through a screen.

"This invention provides for household use a cheap, neat, and handy egg beater, which will be simple in construction, and will beat the eggs more rapidly and thoroughly than egg beaters presently in use."

CHARLES ALDEN (1817-1887)

Figure 2: Portrait of Charles Alden

Charles Alden invented the Alden Process of Pneumatic Evaporation and Super-maturation, a process of desiccating (drying out) meats, vegetables, and fruits. Alden, a native of Randolph, MA, showed a knack at a young age for invention and throughout his life received many patents for his inventions. He was one of the earliest manufacturers of condensed milk and set up the first condensed milk factory in this country, not far from Poughkeepsie, NY. He followed this by establishing a factory in Newburgh, NY for condensing milk and drying meats and vegetables. (Figures 2 and 3)

Alden was considered the leading inventor in all methods of preserving by removing water, leaving only the solid portions. He worked with the Army in developing methods of supplying vegetables and fruits in a form that would keep a long time and could be transported great distances.

The Alden Fruit Preserving Company was in the business of fruit growing and drying. They claimed the Alden Process as the oldest, the best, and the cheapest process for preserving fruits, vegetables, and meats. Alden believed the most essential thing in fruit drying was a rapid current of air through the evaporating chamber.

Over two hundred Alden evaporating factories were in operation in the United States at one point.

Alden was at one time very wealthy and served a number of terms in the New York Board of Aldermen. His untimely death in 1887 was noted in the New York Times headline "Suicide of an Inventor," and was attributed to mental problems due to financial difficulties.

THE

ALDEN PROCESS

OF

PRESERVING AND PERFECTING

Fruits, Vegetables, Meats, Fish,

&c.,

BY PNEUMATIC EVAPORATION AND
SUPER-MATURATION.

ALDEN FRUIT-PRESERVING COMPANY
123 Chambers Street.
NEW YORK.

Figure 3: The Alden Process of Preserving

PRESERVING FRESH FISH, PATENT NO. 235,116
Charles Alden, Gloucester, Massachusetts, December 7, 1880

Alden's new process for preserving fresh codfish and other fresh fish, as an article of food, without the aid of salt or other condiments, using an evaporating pan, a heating coil, steam chamber, air fan, and a movable set of blades.

This patent and the many other patents by Alden created a new industry in the preservation of all types of fruits, vegetables, meats, fish, etc.

JAMES BROWN HERRESHOFF (1834-1930)

James Brown Herreshoff, was the oldest child of Charles Frederick and Julia Ann (Lewis) Herreshoff. He was born March 18, 1834 at Point Pleasant Farm, Bristol, RI. Herreshoff was educated at schools in Bristol and Providence and studied chemistry at Brown University. He was a talented chemist and from 1852 to 1863 was employed by the Rumford Chemical Works in East Providence, RI as a manufacturing chemist.

Herreshoff's inventions included a sliding seat for rowboats (still used in racing shells). Along with his brother, Captain Nathanael, he perfected a new type of steam boiler for light vessels, made in the form of a beehive with coils of iron pipe, which heated water quickly while saving fuel. This invention led to the Herreshoffs' involvement with steam-powered vessels. The coil-type boiler was adopted for the first torpedo boat built for the United States Navy. Herreshoff also developed an engine that could be operated with superheated steam. His other inventions included a fin keel for racing yachts, which by virtue of its depth and weight allowed the yacht to carry unconventionally large sails, the first internal combustion naptha-powered motorcycle in America, mercurial anti-fouling paint, and an apparatus for measuring the heat of gases.

Herreshoff is included here for his invention, in 1874, of a simple press which pressed fish, beets, and other soft food substances that could benefit by compressed processing. (Figures 4 and 5)

PRESSES FOR FISH, BEETS, & c., PATENT NO. 146,338
James B. Herreshoff, Bristol, Rhode Island, January 13, 1874

Herreshoff invented "a new and useful device for pressing beet pulp, fish or any other substance requiring compression." The device is simple in its construction and operation, and at the same time is capable of producing better results more satisfactorily than those that were presently being used.

Figure 4: Herreshoff Boat Works

Figure 5: Herreshoff Mfg. Co.

DEVICE FOR CUTTING CORN FROM THE COB, PATENT NO. 57,361
F. A. Morley, New York, New York, August 21, 1866

This invention cuts green corn from the cob. Simple, efficient and requires less labor.

After an ear of corn is laid onto the bed plate, the user pushes the sliding head forward, forcing one end of the ear into the guides, which picks that end of the ear up and centers it for the knives. The pressure upon the block causes the other end of the ear to be picked up also and centers it.

The knives and their guides can cut the smallest ears, or expand to accommodate the largest ears.

ICE CREAM FREEZER, PATENT NO. 186,438
John F. Rote, Reading, Pennsylvania, January 23, 1877

An ice cream freezer that contains a compound beater, or a beater within a beater. The result is a "thorough agitation and circulation of the cream from side to side to center, and vice versa, by which an evenly frozen mass is obtained, and the formation of coarse particles is entirely prevented."

CUTTING AND PANNING CAKES, PATENT NO. 25,767
John H. Shrote, Baltimore, Maryland, October 11, 1859

This invention cuts dough into cakes and delivers them on a pan for baking.

After the dough is rolled to the thickness required, it is placed on the top of the cutter and rolled over sufficiently hard to cut the cakes. The cakes then fall down onto the pan in the proper position to be placed in the oven for baking. The cutter plate can be raised up on its hinges to allow scraps of excess dough to be removed easily.

CORN POPPER, PATENT NO. 186,279
George P. Sisson, Northhampton, Massachusetts, January 16, 1877

Prior to Sisson's invention, corn poppers had a continuous rotary motion. This corn popper had a reciprocating rotary motion, simply operated by a gear handle connected to an eccentric gear.

The continuous reciprocating rotary motion of the hopper greatly facilitates the popping of the corn without the danger of it becoming injured by scorching or burning.

BUTTER MOLD, PATENT NO. 256,467
Joseph De Carly, Smith's Ranch, California, April 18, 1882

To shape a roll of butter, a strip of cloth as wide as the cylinder is rolled upon the reel, and one end of the cloth is then brought forward to line the cylinder, and the sides hooked on the hooks. A piece of butter is then put into the cylinder on the cloth, and by using the foot lever, the hinged ends of the cylinder are closed. The butter can be molded and clothed in one operation, the butter is firmly pressed, and no handling of the butter is required during the entire process.

MILL FOR GRINDING
TORTILLA, GREEN CORN, & c., PATENT NO. 220,525
Frederick A. Gardner, Buffalo, New York, October 14, 1879

A simple and durable mill that may be conveniently attached either to a vertical or horizontal support, and easily disassembled for cleaning.

PEACH CUTTER, PATENT NO. 203,085
James H. Smith, Monticello, Arkansas, April 30, 1878

A machine for dividing a peach into two parts and removing its stone. Pressing down on the lever arm makes the cutters divide the peach in two, and then the cutters will grasp the peach stone until the next peach is cut.

COMBINED CAN AND MEASURE, PATENT NO. 194,112
Joseph Sears, Chicago, Illinois, August 14, 1877

This lard can serves as a predetermined measure for liquids or solids. The mouth and the cover form a tight-wedging joint. The top flange of the vessel prevents liquids when being poured from running down the outside of the can.

VEGETABLE GRATER, PATENT NO. 98,832
Samuel M. Wilson, New York, New York, January 11, 1870

A rotary grater that is "perforated with small round holes, having an upward inclination, so as to form an overhanging cutting lip, uneven and jagged, while the lower sides of the holes are smooth, and free from jagged points, thus forming a very effective grater, having little liability to clog."

OYSTER OPENING MACHINE, PATENT NO. 189,966
Thomas W. Temple, Los Angeles, California, April 24, 1877

Although an expert with an ordinary oyster knife can easily open one oyster, a large number of oysters is time-consuming and tedious.

This splitting machine opens the oysters "in a very expeditious, simple, and easy manner."

STEAM COOKING APPARATUS, PATENT NO. 174,048
Carl R. Winkler, Toledo, Ohio, February 22, 1876

This steamer has segmented openings, which may be opened or closed to control the passage of steam into the upper vessels, so different kinds of food may be cooked at the same time by the direct or indirect application of steam.

SHELVES FOR GRIDDLES, &C. PATENT NO. 148,580
Sally R. Stevens, Cleveland, Ohio, March 17, 1874

An Improved Cook Stove Companion for holding stove lids or covers, griddles, and other utensils, when not in use on or about the stove.

HOLDERS FOR PLATES WHILE BEING WARMED, PATENT NO. 111,399
William T. Stoutenborough, Brooklyn, New York, January 31, 1871

In households, when it becomes necessary to warm plates for use at the table, the plates are generally placed one upon another in the oven or upon the top of a stove. When warmed in this manner they do not receive the same degree of warmth, and the lower plate, which is in contact with the heated plate of the stove, very frequently cracks and breaks from the heat.

This plate warmer can be set upon the top of a stove, designed so the plates receive an equal degree of warmth, and in which there will be no danger of the plates striking against each other, or falling out, or cracking with too great heat, or by the weight of the plates piled one upon another.

CHAPTER 10

Laundry

"We should all do what, in the long run, gives us joy, even if it is only picking grapes or sorting the laundry."

— E.B. White

During the 1800s, the drudgery of washing clothes was something all families faced, regardless of social status or economic standing. Wash day was tedious, time consuming, labor intensive, and just plain hard work. Women had to heat, carry, and empty all of the water needed for washing and rinsing of clothes. Without running water, and with fire as the only source of heat, this could be quite a daunting task. Laundry took all day. Monday became Laundry Day because many churchgoers took a Saturday night bath and dressed in their Sunday best for church. With ironing taking place the day after Laundry Day, Tuesdays became Ironing Day.

Figure 1: Mrs. Stewart's Bluing

A wash stick was used to stir and whack the clothes. The "dolly-stick" (sometimes known as a "peggy stick") was a wooden stick with a T-handle on the top and four to five wooden fingers or prongs protruding from a disk on the other end. To use it, the operator pumped up and down inside the wash tub and rotated the handle vigorously from side-to-side to agitate the clothes. Laundry washboards were wooden and covered with a corrugated zinc surface on which the clothes were rubbed up and down to scrub them clean. A crank-operated hand wringer then pressed the clothes between two rollers to squeeze out the excess water in them. After boiling, the wash was

One wash, one boiling, and one rinse could end up requiring 50 gallons (about 400 pounds) of water. If soap could not be bought or was considered too expensive, it had to be made. This involved saving and rendering animal fat (lard) and boiling it up with a caustic solution of lye made from wood ash and water. The water for the laundry may have come from a pump nearby, or drawn from a well if you lived in the country, or there may have been a water source in the kitchen or scullery (laundry room). The water was poured into and heated in large copper boilers (known as coppers) on a stove and then ladled into a washtub where the soiled clothes, which had been soaked overnight, were scrubbed with soap on a washboard before going into the refilled copper on the stove. Here they were boiled with washing soda and other cleansing and bleach solutions.

lifted out into clean water to be rinsed. White garments were rinsed and then "blued" for extra whiteness and often starched before being wrung out either by hand or with a mangle or other type of wringer. Bluing (Figure 1) was used to wash linens like towels, bedsheets, handkerchiefs, and table napkins. Bluing was a dye that was added to the wash water to tint white fabrics a light blue that served to counteract any greying in the fabric, returning the linens to a brighter white color. Flat or "sad irons" were placed on the kitchen range to heat. Several irons were often needed at the same time, as they cooled down quickly. Many women would have two, three, even four irons on the stove or fireplace at a time. When one iron would cool, they would put it back on the hearth and pick up another and continue ironing.

While products like the sewing machine made some tasks of homemaking easier, laundry work was not one. Manufactured cloth greatly increased the amount of clothing that people owned, and the availability of washable fabrics, such as cotton, increased the need for washing. Laundry was one of the first chores that women, at least those who could afford it, were able to pay others to do. Even women of limited means sought relief in the form of washer women, commercial laundries, and mechanical aids.

The earliest washing machines used a lever to move one curved surface over another, rubbing clothes between two surfaces, each surface ribbed like a washboard. This early device basically imitated the washing motions of human hands.

Of the over 1,700 patents issued for washing machines during the 19th century, most were never produced. The majority that were produced were for commercial laundries. Washing machines began to appear in homes immediately following the Civil War, but didn't catch on in great numbers until reliable sources of water and power became available.

The 1876 Centennial Exhibition in Philadelphia and the 1893 Columbian Exposition in Chicago both had Women's Buildings that featured exhibitions of inventions by women. Although women had exhibited at previous American fairs, the 1876 Centennial marked the first time that there was a separate building devoted exclusively to products and inventions by women. At the Women's Pavilion at the 1876 Centennial, most of the women's inventions pertained to the household. Many of these were designs to make household work easier. At least half a dozen women showed laundering and ironing devices.

Sybilla Masters is often named as the first woman inventor, for a method of making cornmeal from maize, although the patent was actually assigned to her husband, Thomas. It was awarded in 1715 by the English courts. Thomas Masters actually received two patents for ideas that originated with his wife.

In 1809, Mary Dixon Kies received the first U. S. patent issued to a woman in her own name. Kies, from CT, invented a process for weaving straw with silk or thread to make bonnets. Her invention was used in the New England hat manufacturing industry.

Martha Coston improved upon her husband's idea for colored signal flares after his death. In 1871, she patented the flare system that was used by the Navy in the Civil War. She sold the rights to the flare patent to the government for $20,000, a very large sum of money at that time.

Sarah Goode was the first African American woman to receive a patent. Born into slavery, she was freed at the end of the Civil War, moved to Chicago, and opened a furniture store. Her invention was for a cabinet bed. She noticed the lack of space in city apartments and invented a foldable bed that, when not in use, could be used as a desk.

From 1790 until 1850, women received approximately 28 patents. During the next two decades, women received at least 350 patents. Women invented what they knew best, around the problems they faced, and so many inventions centered around the home. With the coming of the Civil War, this gradually began to broaden. The war forced many women into new roles and experiences. With men away, women worked in factories and on farms. Increased educational opportunities in the second half of the 19th century also contributed to the increase in the number of inventions by women.

The Patent Act of 1790 offered women the same patenting privileges as men. However, laws in most states prohibited married women from owning property, including inventions, in their own names. Married women in the early 19th century had no legal rights. In most cases, a woman's husband controlled all of her property. Inventing and obtaining a patent also took time and money. During this time, many women's inventions were patented under the names of their fathers, husbands, brothers, even sons. Some women only used their first initials on the patent papers. The number of patents issued to women increased with changes in property rights law. Removal of legal barriers not only gave married women patentees the right to sell patent rights without spousal approval, but also allowed them to keep their own earnings. Even with these advancements, by 1910, inventions patented by women accounted for less than 1% of all patents issued in the U.S.

In the washing machine industry, production centered in the midwest, the heartland of heavy industry. It is here, in Iowa, that J.B. Woolsey received a patent for his improved washing machine. Woolsey's washing machine is noteworthy for incorporating several features into one machine. It was designed to agitate the clothes and had a flue which could be hooked up to the chimney in the house to vent the machine. It may also have been the first washing machines that could heat the water right inside the tub. Woolsey had the foresight to indicate in the claims of his patent that it was suitable, not only for home use, but also for larger laundry facilities. Woolsey's model is noteworthy today for its folk art quality and simplicity of line. The inventor used existing technology, considered the needs of the time, and adapted his invention to simplify the clothes washing process for both commercial and home use. His additions—the crank for agitating the wash, a fire box to heat the water, and a flue for ventilation are features that are still found in washing machines in use today.

WASHING MACHINE, PATENT NO. 90,416
J. B. Woolsey, Bloomfield, Iowa, May 25, 1869

Woolsey's patent was one of the very first washing machines to actually heat the water in the wash tub. To operate the washing machine, the tub was filled with a sufficient quantity of water and a fire was built in the fire box, which heated the water to the desired temperature. The clothes to be washed were put into the cylindrical tub along with the soap or other cleansing material. The tub was rotated by a crank handle causing the clothes to be rolled around in the tub and rapidly cleansed by the water passing in and out of the flanged openings built into the tub. The fire box could be used for either wood or coal and the smoke pipe would be inserted into the smoke flue of the chimney in the house. If you added a motor to rotate the cylindrical tub and replaced the fire box with a heating element, Woolsey's Washing Machine from 1869 would bear a close resemblance to today's modern washing machine.

JACOB BRINKERHOFF (1834-1913)

Jacob Brinkerhoff was a prolific inventor who received over two dozen patents encompassing many different subjects ranging from corn shellers to wire fence nails. Several of his inventions were for improvements in clothes wringers.

Brinkerhoff, along with several partners, formed the Empire Clothes Wringer Company in Auburn, NY in 1872. The company was formed for the purpose of manufacturing the clothes wringer from Jacob Brinkerhoff's patent. The Empire Wringer Company had the exclusive control of the patents. The business moved locations several times in its history, always staying in the Auburn area. In 1876, 26,000 wringers were sold and in this same year the company also started producing folding cots and washing benches. In 1881 Jacob Brinkerhoff was elected President of the company. In 1882, 60 people were employed at Empire Wringer and produced over 100 wringers per day. (Figure 2)

Figure 2: Advertisement Card, Empire Wringer Co.

CLOTHES WRINGER, PATENT NO. 146,867
Jacob Brinkerhoff, Auburn, New York, January 27, 1874

This invention combines a steel spring with a stop to preserve the spring from becoming broken. Unless a steel spring is protected when used in wringers, it is liable to get overstrained and break. Brinkerhoff overcomes this issue by controlling the tension on the spring by the use of stops that prevent unwanted strain from developing.

WASHING MACHINE, PATENT NO. 106,137
Charles H. De Knight, Pittsburg, Pennsylvania, August 9, 1870

The rubbing disk within the wash tub, the lower surface of the disk, and the upper surface of the bottom of the the tub had a series of curved rubbing strips. An adjustable spiral spring pivots the disk to a vertical shaft, causing "a rotating reciprocating motion."

SAD IRON HOLDER, PATENT NO. 158,841
Mary E. Hildreth, Mount Pleasant, Iowa, January 19, 1875

A device to keep an iron pressed flat on the fabric. Or, in other words, "a means for holding, against endwise displacement, a guard holder on the handle of a sad iron when the sad iron is passed, in the act of ironing, to and fro over a garment or other article. It also prevents the hand of the operator from being brought into contact with the heated handle of the sad iron."

A sad iron was not melancholy, but merely heavy.

WASHING MACHINE, PATENT NO. 110,825
Sarah J. Clark, Richmond, Indiana, January 10, 1871

This simple washing machine supported a wash board in its tub, so the operator could brush the soaped clothes on the washboard. The upper surface of the washboard is covered with a corrugated zinc plate and a valve is located under the tub to drain the wash water.

SAD IRON, PATENT NO. 137,201
John Hewitt, Pittsburgh, Pennsylvania, March 25, 1873

This iron combines both sad (heavy) iron and polishing iron features. The polishing face is at or near the center and imparts a gloss to the shirt-bosoms.

IRONING BOARD, PATENT NO. 181,034
Mary A. Bryant, Freeport, Maine, August 15, 1876

This adjustable ironing board could provide surfaces specific to the ironing of shirts, skirts, and other articles which require the board to be inserted within them.

SAD IRON HOLDER, PATENT NO. 294,196
Serena M. Carnes, New York, New York, February 26, 1884

This invention improved on Carnes' Patent No. 269,006 dated December 12, 1882. The clasp spring of the holder now allows circulation of air to the underside of the pad, keeping it cool.

WRINGING MACHINE, PATENT NO. 226,647
Friedrich Bernhardt, Fischendorf, Kingdom of Saxony, April 30, 1879

This machine is especially constructed for "pressing out from endless fabrics, the mixture of water and soap, or water, soap, and grease, with which they may contain, for the purpose of recovering said ingredients in a concentrated state."

This Wringing Machine was also patented in England on April 30, 1879.

IRONING TABLE AND CLOTHES RACK, PATENT NO. 156,749
William Curtis Arnold, Montague, Michigan, November 10, 1874

This invention neatly combines ironing tables, storage, and clothes racks or driers that are arranged to be folded in a compact form, and composed of a table frame, from which the main rack standards rise, and to which standards are hinged additional racks. The table box has receptacles for clothes, pins, and sad irons, and is constructed with three ironing tables and three clothes racks. The main board of the box table is provided with a hinged cover for the reception of the clothes to be ironed and has stationary sad iron rests.

CLOTHES DRIER, PATENT NO. 11,985
Stephen Woodward, Sutton, New Hampshire, November 21, 1854

Most clothes driers were arranged so an adult could easily reach them to place and remove clothes. Woodward's improved clothes drier could be lifted entirely above a person's head allowing one to pass under the clothes being dried. The raised height of the clothes drier also allows for the clothes to be dried faster.

FUEL BOX AND WASHING APPARATUS WITH SETTEES, PATENT NO. 35,210
Martin Bishop, Washington, District of Columbia, May 13, 1862

Bishop has invented a very unusual combination of a Settee, Fuel Repository, and a Wash Stand. The settee is constructed with a front panel, a bottom, and a back. The settee is divided into three separate compartments. The compartment at the right hand side is lined with sheet metal for the storage of coal. The middle compartment is for wood or kindling and the left hand side is for the wash stand. The seat forms the lid to cover these different compartments.

CLOTHES POUNDER, PATENT NO. 224,747
John W. Tullis, Fairfield, Illinois, February 17, 1880

This device improves on cone or cup shaped washers that employ a spring for yielding upward when pressing down upon the clothes and preventing the clothes from rushing up too far in the cone or cup when it is forced down upon the clothes.

The advantages of this improved clothes pounder are that, while the clothes are prevented from being forced too far up in the cone or cup, the small spiral or coiled springs rub the clothes lightly, thus softening and loosening the dirt both in their yielding or vibrating action, when the pounder is forced down upon the clothes.

WASHING MACHINE, PATENT NO. 185,274
Samuel C. Wilson, Forest City, Arkansas, December 12, 1976

This complex washing machine combines a wash tub and an upright frame. The frame supports bearings and a shaft, producing power to drive the machine as the handle is cranked.

CLOTHES MANGLE, PATENT NO. 216,320
James S. Fox, Oshawa, Ontario, Canada, June 10, 1879

This invention connects a mangle to the folding table, so it can readily be removed so the table then be used as a conventional ironing table or bench. The ironing table also stores skirt and shirt boards.

WASH BOILER, PATENT NO. 214,782
Adelaide E. Mann, Lawrence, Massachusetts, April 29, 1879

Adelaide Mann had several patents for washing apparatus and women's garments. This patent provided for a basket with a hinged bottom, suspended above the bottom of the wash boiler. "The articles to be boiled or cleansed are placed in the basket and a sufficient amount of water is added to the tub. The bottom and the sides of the basket are a short distance away from the bottom and the sides of the boiler, thus preventing the clothes from coming into contact with the boiler. When the boiling process is completed the basket is raised up out of the water so as to allow the water to drain out of the articles."

CHAPTER 11

Leather & Shoes

"The shoe that fits one person pinches another; there is no recipe for living that suits all cases."

— Carl Jung

The Industrial Revolution brought leather tanning and shoemaking from villages, where skilled craftsmen worked in their homes or farms, to the city, where workers labored away in large factories.

Tanning is the process of treating skins of animals to produce leather, using either vegetable or mineral tanning agents. Vegetable tanning uses tannin. Tannin occurs naturally in bark, with hemlock a preferred source. Tannins bind to the collagen proteins in the hide, coating them, and causing them to become less water soluble and more resistant to bacterial attack. The process also causes the hide to become more flexible. Vegetable-tanned hide is currently used for shoe soles, luggage, saddlery, belts, and some upholstery. Mineral tanning usually uses chromium. Chrome tanning, which came later, is faster than vegetable tanning and produces a stretchable leather used in handbags and garments. (Figure 1)

In the United States, Colonial farmers frequently tanned their own leather and made their own shoes. Later, itinerant cobblers traveled from town to town, making very basic, silver-buckled shoes that could be worn on either foot. (Figure 2)

The tools and processes of shoemaking before 1830 were few and easily mastered. A shoe had only two major parts: an upper and a sole, and only four processes were required: cutting, fitting, lasting, and bottoming.

Only eight hand tools were used. The cobbler used a

Figure 1: Leather Tanning Process

single knife to cut both the sole and upper leather. Next, he used a lapstone and a handmade hammer for pounding the leather. To prepare for sewing, he used an awl to bore holes, and a needle or a bunch of bristles to sew the shoe. The process of sewing was called fitting, as he stitched the parts of the upper together to the sole. After the upper was fitted, it was slipped onto the last, the form that shapes the shoe while it's being made, which had an insole tacked to it. The lower edge of the upper was pulled over the wooden form tightly with pincers, until it could be fastened temporarily with nails. Then the outer sole was either sewed or pegged to the lasted upper. The last in the shoe was held firmly in place by a strap or stirrup, which passed over it and down between the shoemaker's knees where the shoe rested and was held taut under his left foot.

The rise of the Central Shops, also called manufactories, marked the beginning of the factory stage of shoe production, this period gaining its most momentum around 1820. Men and women shoemakers, working out of their homes, received shoe material and instructions from the Central Shop, they completed the shoes, and then sent the finished shoes back to the shop.

Carriers delivered the material for several dozen shoes to each house, first to have the uppers stitched and bound, and then, often at another house, to be lasted and soled. Men, most of them part-time shoemakers and part-time farmers, worked independently, and were paid by the piece. The backyard workshops where the shoes were made were called "ten-footers" as they were usually 10' x 10'. Initially, each man made a complete shoe, except for cutting the leather and the final binding. Women also worked, not in the ten-footers, but in their kitchens, where they bound the shoes. They were not paid for their work separate from their husband's pay until the early part of the 19th century, when shoemakers began hiring apprentices and assistants to work with them, and women started working directly for the shoe boss, as he was called.

The 1830s and 1840s saw a number of changes in the shoe industry brought about by increasing competition, a growing market, and a greater demand for stylish, high quality shoes.

From the mid 1750s through the 1920s, Massachusetts was a center for shoe manufacturing. Many immigrants moved to the area as they were already familiar with the shoemaking craft as practiced in their native countries. The manufacture of leather goods, especially boots and shoes, was one of the leading industries in New England during the 1800s and early 1900s.

The Massachusetts towns where the shoe manufacturing took place became factory cities with specific towns known for particular areas of expertise. Lynn specialized in women's and children's shoes, along with many shoe-related inventions. Natick and Danvers were recognized for heavier, coarser shoes called brogans. Randolph was noted for its fine quality boots.

Industrialization changed the work of the shoemaker during the mid-1800s. Initially, the work done in the factories was identical to the work done in homes. However, the need for standardization of products contributed to the development of machinery in the boot and shoe industry. Technological improvements included both new tools and techniques for cutting and sewing leather, which increased the uniformity of the product along with new patterns and styles of shoes. Tin patterns to guide the cutting of the sole insured greater accuracy. Marketing innovations increased standardization, including a new system of sizing which

Figure 2: Shoe Cobbler at work c. 1880

allowed a better fit and the invention of the "crooked shoe," a different shoe for the right and left foot. The shoe box was also invented during this period.

New machines made the manufacturing of shoes cheaper, quicker, and more uniform. In the late 1850s, a variety of new shoemaking machines were invented. New methods of cutting, finishing, and packaging of shoes appeared. Laced shoes as well as buttoned shoes became the fashion. Strong competition among shoe manufacturers pushed the development of new ways to make shoes. The invention of these machines revolutionized the shoe and boot industry, greatly increasing the number of shoes produced and improving their quality.

Elias Howe's 1846 invention, the sewing machine, was quickly adapted for use in stitching the upper parts of shoes together. In 1858, the master mechanic, Lyman R. Blake, patented a machine for sewing the uppers of shoes to the soles. Gordon McKay of Pittsfield, MA bought Blake's invention for $8,000 in cash and a share of future profits, and patented an improved version of the machine in 1862, the year he received an order from the government for 25,000 pairs of cheap, sturdy boots. From 1862 to 1890, McKay (alone or with others) filed over 40 patents for sewing, nailing, tacking, lasting, and pegging machines for the mass production of shoes. McKay recognized that his fortune lay not with shoe production but with the machinery that was needed to produce the shoes. He did not sell his sewing machines outright; instead he leased them to manufacturers, and developed a leasing system for them based on selling royalty-stamps that kept track of the shoe pieces made. The McKay sewing machine became one of the leading sewing machines used in the United States and Great Britain, with the shoes made on this machine coming to be called "McKays." McKay's machines transformed the manufacture of shoes, and by the late 1870s, they had been responsible for producing 120 million pairs, half the nation's shoes.

Charles Goodyear Jr. patented the Goodyear Welt sewing machine in 1871. It could quickly stitch together the welt, upper, and soles without any hand-sewing. Goodyear was the son of Charles Goodyear, the famous inventor of the process of vulcanizing rubber. Goodyear Welted construction got its strength, durability, and repair qualities by stitching the upper leather, lining leather, and welt (a specially prepared piece of leather that joins the upper to the sole) to ribbing that was already bonded to the insole. The welt was then stitched to the leather or rubber sole. Shoes and boots using the Goodyear Welt construction could be resoled repeatedly. (Figure 3)

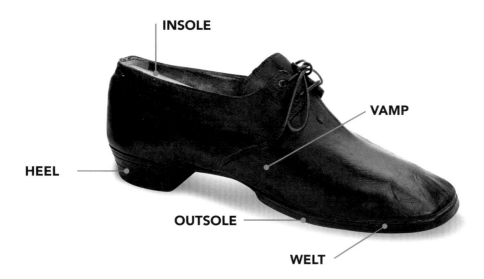

INSOLE

VAMP

HEEL

OUTSOLE

WELT

Figure 3

LEATHER CREASING MACHINE, PATENT NO. 178,199
Joseph B. Stetson, Lincoln, Maine, April 29, 1876

For harness making, this device produced leather straps of various widths and thicknesses "rapidly, easily and with uniformity," to add ornamentation that imitated handwork.

SOLE BURNISHING MACHINE, PATENT NO. 130,567
Augustus C. Carey, Malden, Massachusetts, August 20, 1872

This machine is an improvement of Carey's earlier patent No. 124,479 which was intended solely for burnishing the shanks of soles of shoes. Additionally this machine is designed to effect the "fore-part" burnishing of the sole that encompasses the narrow band about the edge of the sole bottom that was common to certain types of shoes and boots of the time.

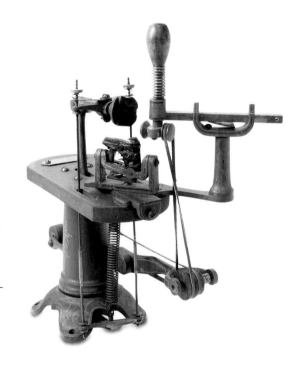

BOOTS AND SHOES, PATENT NO. 169,006
Emanuel S. Justh, Washington, District of Columbia, October 19, 1875

These metallic plates stamped out of sheet metal extend the life of the shoes' soles. They are attached to the soles by prongs, at points where the soles are most subject to wear.

MACHINERY FOR GORING SHOE UPPERS, PATENT NO. 141,642
Asahel J. Goodwin, Brookline, Massachusetts, April 11, 1873

Goring is an elasticized material that joins the two upper pieces of a shoe which provides for a more flexible opening and can eliminate the need for laces. This machine is an improvement and more convenient and useful than Goodwin's Patent No. 133,703, dated December 10, 1872.

LEATHER SPLITTING MACHINE, PATENT NO. 74,734
Francis J. Vittum, Newburyport, Massachusetts, February 18, 1868

Starting around 1830, cattle hides were split by machines that produced both stout and light grain leather, with the best splits used for boot backs and shoe quarters. Vittum's "endless belt splitting machine" is a complicated structure that split the leather more precisely than past machines.

NAILING LASTS FOR BOOTS AND SHOES, PATENT NO. 131,565
Joseph A. Safford, Winchester, Massachusetts,
American Boot And Shoe Machine Works, September 24, 1872

The last holds a boot or shoe firmly in place while the shank is being nailed or the bottom is finished. This last is hinged together at the toe, with the sole portion pivoted at the heel end to the adjustable standard and the instep portion connected to the standard by pins working in a slot, or in cam grooves.

MACHINE FOR CUTTING SHANK STIFFENERS, PATENT NO. 170,135
Jeremiah M. Watson, Sharon, Massachusetts, October 28, 1875

This device cuts shank stiffeners for the soles of boots or shoes using two reciprocating knives. When the shaft revolves, each knife not only moves toward and away from the bed, but also crosses the path of the other knife, so the leather is cut first by one knife and next by the other, beveling the shank stiffener's edges.

LASTS FOR BOOTS AND SHOES, PATENT NO. 189,418
John Batley of Kensington Park Gardens, John Keats of Wood Green, James Neil of Worship Street, England, April 10, 1877

Adjustable lasts for the manufacture and display of boots and shoes provides for the contraction and expansion of the last at the heel, or between the heel and the instep, and at the sides of the forward portion of the foot, to facilitate its entry within a boot or shoe.

LASTING MACHINE, PATENT NO. 217,865
Charles W. Gladden, Lynn, Massachusetts, July 29, 1879

The upper holding jaws of the last are automatically but independently closed, to adapt to materials of different thickness. It was essential to hold the upper firmly in place to prevent slippage during lasting.

BOOT AND SHOE, PATENT NO. 238,534

Lewis Slessinger, San Francisco, California, March 8, 1881

The vamp is the part of a boot or shoes that covers the tops of the toes. In this boot design, the lower part of the seam is formed two or three inches above the quarter and vamp by overlapping the edges of the front and back parts, and uniting them with rivets and stitching.

TRIMMING OF SOLES OF BOOTS & SHOES, PATENT NO. 136,790

Benjamin J. Tayman, Philadelphia, Pennsylvania, March 11, 1873

This machine trims shoe soles and welts and also burnishes the soles. The revolving shaft has cutters and a burnishing wheel, and may be oriented in either a horizontal or vertical position even while the machine is running, thus preventing loss of time in making the adjustment.

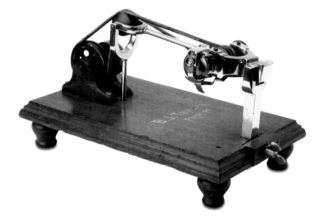

SOLE EDGE BURNISHING MACHINE, PATENT NO. 173,284

Charles H. Helms, Poughkeepsie, New York, February 8, 1876

The burnishing machine is used for setting or burnishing the edges of the soles of boots and shoes. The tool is steam-heated and features rests for the shoe and the operator's hand, to ensure steady contact between the sole and the tool.

TANNING MACHINE, PATENT NO. 106,085
John H. Slocum, George F. Turner,
Fayette, Maine, August 2, 1870

This tanning reel allows hides and skins to be evenly tanned into leather. The cylindrical reel has removable slats with hooks for attaching hides. As the operating gear revolves the reel in a vat filled with "the proper liquor," the hides tan in about one third of the time required by the old process.

MACHINE FOR LEATHERING TACKS, PATENT NO. 20,819
Jesse Reed, Marshfield, Massachusetts, July 6, 1859

The operator feeds tacks and leather into the machine, which punches the tack head through the leather, and then drops the leathered tack out of the machine.

BOOT AND SHOE CLEANER, PATENT NO. 210,072
G. Frederick Ziegler, Jersey City,
New Jersey, November 19, 1878

The brushes for this shoe cleaner are contained in a pan. The weight of the user keeps the cleaner in place while in use, and the dust or dirt received into the pan can be emptied out whenever desired.

BOOT & SHOE, PATENT NO. 231,632
Sarah A. Thecker, Georgetown, District of Columbia, August 24, 1880

An improved means for fastening ladies' and children's button shoes, whereby a uniform pressure upon the instep is secured, a neat finish is obtained, and the use of shoe buttoners is dispensed with.

CHAPTER 12

Manufacturing

"Manufacturing is more than just putting parts together. It's coming up with ideas, testing principles and perfecting the engineering, as well as final assembly."

— James Dyson

Before the American Revolution, English law mandated that the colonies trade only with England. Tough laws prohibited the export of machinery, plans, models of machines, even the mechanics themselves. England obstructed growth of American industry by imposing custom duties favoring English products and domestic taxes on incoming American products. As a result, much of American industry flourished after the Revolutionary War ended in 1783. After the revolution, the U.S. government raised custom taxes against England to protect its growing industries.

Manufacturing in the U.S. accelerated with the growth of factories, mass production, and the increasing use of interchangeable parts. From mainly workshop or home manufacture, in the 19th century, American industry moved to the factory. The textile industry was one of the first to effect this migration, as spinning and weaving of silk, wool, cotton, and flax were gradually mechanized. Mechanization increased the quantity and quality of production. The introduction of new types of machines also contributed to the growth of manufacturing. Steam power began to replace horse, water, and manpower.

The growth in the textile and clothing industries was directly attributable to the invention of the sewing machine. The boot, shoe, and textile industries were all examples of small shops eventually being replaced by large factories. The introduction of power-driven sewing machines in the manufacture of military uniforms, boots, and shoes brought these industries into the factory system. Before the Civil War, the U.S. was chiefly an agricultural nation. This changed with the introduction of mass production, the use of standard, interchangeable parts, and increasingly efficient methods for parts assembly.

The first U.S. textile factory was built by Samuel Slater in 1790. Slater built a cotton spinning mill in Pawtucket, RI which was run by water power. As a result, Andrew Jackson dubbed him the "Father of the American Industrial Revolution."

In 1780, Oliver Evans invented an automated flour mill that would eventually replace the traditional gristmill. Evans' automated mill did not require human labor. In 1804, Evans also developed one of the first high-pressure steam engines and began establishing machine shops that could manufacture and repair his inventions. Evans' steam engine could be used for a variety of industrial purposes. Within a few years, it was being used to power ships, sawmills, flour mills, and printing presses, in addition to textile factories.

During this period, trade throughout the U.S. expanded. Small farms waned and more consumer goods began to find their way into the marketplace. The transition away from an agriculture-based economy towards machine-based manufacturing led to a great influx of people from the country moving into the cities in search of work, causing the population of towns and cities to increase.

Throughout the 1800s, factory production grew in the U.S., spreading beyond the New England textile mills to other regions and industries. These industries grew and flourished due to a number of factors. One was plentiful, natural resources. America had great amounts of forest, plentiful water, and vast mineral wealth, including coal, iron, copper, silver, and gold. Industries used these resources to manufacture a wide variety of goods. Another key factor was improved transportation, as the development of steamboats, canals, and railroads made it possible to ship items long distances more quickly. Railroad building boomed after the

Civil War. As shipping raw materials and finished goods became easier, industry grew in its wake. The final factor in this equation was power. Power sources to operate machinery increased in number and became more sophisticated, and capable of supplying cheap and reliable energy. Steam power made it possible to establish shops far from waterways and closer to sources of raw materials and labor. The steam engine was rapidly adapted for use in all kinds of manufacturing.

As discussed in Chapter 3, Alcohol, Tobacco and Firearms (ATF), Eli Whitney was responsible for implementing the idea of standardized, interchangeable parts. He entered into a contract to make muskets for the government, attempting to meet the army's goal of interchangeability, to make it possible to repair small arms in the field without requiring skilled gunsmiths or specialized tools. Whitney is best known for inventing the cotton gin, which separated the cotton seeds from the raw cotton fibers (1794, patent # X72). This invention, combined with applying steam power to the manufacture of cotton goods, contributed to increasing the role of cotton production in the South. The cotton gin, the railroads, and slavery all combined to make cotton planting a very profitable business.

The Industrial Revolution was the major technological, socioeconomic, and cultural driver of the late 18th and early 19th century, as an economy previously based on manual labor morphed into one driven by industry and machine-driven manufacture. The two most notable features of manufacturing were extensive use of interchangeable parts and extensive use of mechanization to produce them. This resulted in more efficient use of labor compared to earlier hand methods.

NELSON GOODYEAR (1811-1852)

Nelson Goodyear was Charles Goodyear's youngest brother. Charles Goodyear invented techniques for manufacturing soft, flexible rubber. In 1839, he invented a process of mixing natural rubber with sulphur to make a stronger and more resilient product; the process was later termed "vulcanization." He patented the vulcanization process for rubber in 1844.

Hard rubber was invented by Nelson Goodyear (patent # 8075). Nelson stated that he used 25 parts or more of sulphur to 100 parts of rubber gum to make a product that was hard and stiff. Ordinarily differences in proportion do not amount to an invention and will not sustain a patent, but the invention of hard rubber was an exception to the rule, for by increasing the amount of sulphur in the composition, an entirely new product is formed having different characteristics than soft, elastic rubber. Goodyear's composition was therefore new and patentable. In 1851 Nelson Goodyear patented hard rubber or Ebonite, also known as vulcanite. Nelson

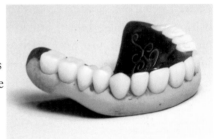

Figure 1: Vulcanite Denture with Porcelain Teeth c.1880

Goodyear received a number of patents for rubber items, but is best known as the inventor of hard rubber. (Figure 1)

MANUFACTURE OF INDIA RUBBER, PATENT NO. 8,075
Nelson Goodyear, New York, New York, May 6,1851

Nelson Goodyear patented and submitted this specimen to showcase a new hard and stiff form of vulcanized rubber. He mixed caoutchouc, also known as India or natural rubber, with substances such as sulfur, magnesia, and lime and then used the heating and curing process previously developed by his older brother, Charles Goodyear. This new form of hardened vulcanized rubber was the starting point of what eventually became plastic.

STEPHEN F. WHITMAN (1823-1888)

In 1843, at the age of nineteen, a young Quaker named Stephen Whitman, set up a small "confectionery and fruiterer shoppe" close to the waterfront in Philadelphia, PA, trying to compete with the fine European chocolates of the time. He drew customers from all walks of life, but it was the sailors from the nearby wharf who provided Whitman with the imported fruits, nuts, cocoa, and flavorings from their travels. He incorporated all of these delicacies into his chocolate.

Figure 2: Whitman's Sampler Box

It was not only Whitman's high quality ingredients and excellent product that fueled his success. Whitman was an excellent marketer. It was his presentation of the candy, as well as its taste, that kept customers coming back. He created beautiful packaging and well-crafted advertising campaigns for his product. His first newspaper advertisement appeared in 1860, just prior to the start of the Civil War. Whitman developed great name recognition through the ads he placed in newspapers and magazines. Even today, his distinctive yellow box with the words Whitman's Sampler and the cross-stitch design is easily recognizable. Whitman Candies is today owned by Russell Stover Candies, Inc.

Whitman began his company selling individual chocolates, but in 1854, he introduced his first box of candies—"Whitman's Choice Mixed Sugarplums" from Stephen F. Whitman. Packaged in a frilly pink container adorned with rosebuds and curlicues, it became the first packaged confection in a printed trademarked package. Sugarplums, at the time, referred to any candy, not just one type.

In 1866, requiring more space for his expanding candy-making business, Whitman moved to a larger, more central location in Philadelphia and began to wholesale his product to other merchants in the Philadelphia area.

Whitman was a businessman, an inventor, and a marketer. He invented the concept of prepackaged chocolate. Whitman's success was due as much to his high-quality chocolates as to his innovative marketing sense. Not content with a small candy shop, Whitman expanded his business, even going so far as to patent the machine he designed to improve his candy production. And as a testament to him, 160 years later, the Whitman name is still today synonymous with chocolate. (Figure 2)

MACHINE FOR CASTING CONFECTIONERY, PATENT NO. 169,935

Stephen F. Whitman, Philadelphia, Pennsylvania, November 16, 1875

This machine for manufacturing confectionery used a steam-filled box surrounding the body to heat the pre-molded batch which was to be poured into multiple vessels located on the top. The underside had funnel shaped apertures where different types of molds could be attached to cast the confectionery. A lever controlled a sliding gate underneath to control several streams of material simultaneously.

U.S. SENATOR JOHN PERCIVAL JONES (1829-1912)

John Percival Jones was an American inventor and politician who served for 30 years as a Republican U.S. Senator from Nevada. He made a fortune in silver mining, was co-founder of the town of Santa Monica, CA, and invented a method for improving ice making.

Born in Herefordshire, England, Jones immigrated to the U.S. with his parents. He attended public schools in Cleveland, OH. In 1849, Jones went to California to join in the Gold Rush. He settled in Trinity County, CA where he worked in mining and farming. He went on to become sheriff of Trinity County, keeping order in the many mining camps. In 1867, Jones was the nominee of the Republican party for Lieutenant Governor. In 1868, he moved to Gold Hill, NV where he was superintendent of

Figure 3: J.P. Jones Residence — Miramar

the Crown Point silver mine which was part of the Comstock Lode. When a body of silver ore was struck in 1870, Jones was one of the individuals who acquired shares, and along with other investors, was able to gain control of the Crown Point mine. He was a member of the California State senate from 1863-1867, was elected to the U.S. Senate in 1873 from Nevada serving 5 terms from 1873 - 1903.

Jones built the first railroad (Los Angles and Independence Railroad) from Los Angeles to Santa Monica and was one of the pioneers of California and Nevada. Jones built the historic residence, Miramar in Santa Monica where he retired in 1903. Subsequently, it was sold to King Gillette, founder of Gillette Razor Company, and is today the Fairmont Miramar Hotel. (Figure 3)

CONGEALING PLATES FOR ICE MACHINE, PATENT NO. 202,265
John P. Jones, Gold Hill, Nevada, April 9, 1878

Jones improved the section of an ice machine known as the congealer, where the ice was formed upon freezing plates. His apparatus made it easier to remove ice from the freezing plates by quickly circulating hot fluid through piping to release the ice and then allow the freezing process to resume.

LATHE FOR TURNING SHAFTING, PATENT NO. 58,605
J.J. Conley, New York, New York, October 9, 1866

This metal turning lathe improved accuracy, repeatability, and safety. Its block system better supported the shaft as it turned, its cutter head had better design and placement, and it used two sets of grooved feed rollers to push and to draw the shaft during operation.

MACHINE FOR CUTTING DIAMONDS, PATENT NO. 216,955
Anthony Hessels, New York, New York, July 1, 1879

Cutting diamonds to shape by hand was a slow and tedious process. This machine was designed to cut facets more accurately by improving the positioning at different angles of inclination. The cutting diamond was cemented to a horizontal reciprocating socket frame jig and it had a lateral vertical adjustable reciprocating carriage to support the jig that held the diamond to be cut. The inventor stated that every diamond received the "gem cut" which was the most perfect and regular of cuts.

MACHINE FOR SAWING SHINGLES, PATENT NO. 240,048
Henry F. Snyder, Williamsport, Pennsylvania, April 12, 1881

When cutting wood shingles, it was desirable to get the most out of a "bolt" of wood. In this machine, a traversing carriage with a pair of spurred rollers seized the wood and fed out a proper thickness before presenting it to the saw blade. Grooves in the spurred rollers allowed stationary "L" shaped guide blades to guide the very tail end of a bolt of wood, thus minimizing waste. This machine was produced commercially.

BRUSH MAKING MACHINE, PATENT NO. 195,017
Thomas Jesson & Thomas Duggan, Galway, Ireland, Glasnevin, Ireland, September 11, 1877

The operator set the "brush head," which was the shell that received the bristles, into a hollow box placed in a recess at the front of the machine bed. The bristle bundle was then laid across the box and a specialized wire clip was set across the center of the clip, forcing down and folding the bristles into the "brush head". Rivets then passed through the "brush head" to hold everything together.

MACHINE FOR THREADING WOOD SCREWS, PATENT NO. 166,121
Benjamin A. Mason, New York, New York, July 27, 1875

Mason's machine manufactured gimlet pointed screws. A rotating screw driver revolved around the screw blank, and was acted upon by revolving milling tools to cut the thread on the screw. This new process simplified the threading process and reduced the force needed.

BRICK MACHINE, PATENT NO. 29,171
James Hotchkiss, Yellow Springs, Ohio, December 5, 1860

This brick machine introduced spiral wings set inside the revolving hopper to wedge the mixed clay into the molds. It also included a strike to smooth the top portion of the brick and a mold wheel that brought the molded bricks up so they could be mechanically lifted up out of the molds and pushed onto a waiting conveyor.

CRACKER MAKING MACHINE, PATENT NO. 71,420
David Stewart, Philadelphia, Pennsylvania, November 26, 1867

This cracker maker sent flattened dough along rollers to be cut into strips and then to length by a knife. Forks then lifted them to a second station to be dusted with flour and stamped before being sent to the oven. The inventor stated, "My machine will therefore make a cracker more uniform, more regularly shaped, more perfect, more pleasing."

MANUFACTURING OF ILLUMINATING GAS, PATENT NO. 140,264
Charles Gearing, Pittsburg, Pennsylvania, June 24, 1873

This gas machine super-heated steam and air, which then carbonized when it came into contact with hydrocarbon liquids. To better dry and superheat the steam and air before making contact with the hydrocarbons, it first traversed a series of pipes and retorts filled with pebbles. After making contact with the hydrocarbons, the vaporized oil was converted into permanent or fixed illuminating gas to be fed to a gas boiler.

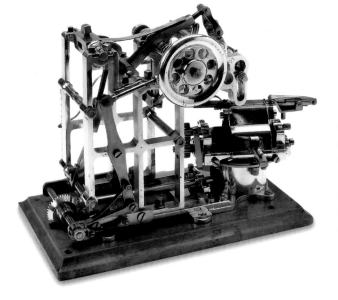

MACHINE FOR CUTTING FILES, PATENT NO. 44,633
James Jervis, Baltimore, Maryland, October 11, 1864

The Jervis machine enabled the workman to cut files with greater accuracy. The file blank was secured to an anvil, that could revolve horizontally or spin on its base as well as move at right angles. The spring-loaded chisel that struck the blank to make the cut could cut at an angle if desired, and the force of its blow could be controlled.

LINK MACHINE, PATENT NO. 233,768
Henry Jerome, Cleveland, Ohio, October 26, 1879

This machine made elliptical links for coupling railroad cars or similar purposes. A shaft and cog wheel act on the first station on the machine, closing the jaws that bend the metal over the die. Once bent, the link can be welded. Finally, the bent piece is flipped and set in the second station for final welding.

PROCESS FOR MAKING TRANSPARENT WINDOW SHADES, PATENT NO. 14,457
Edward R. Kernan, Pittsburgh, Pennsylvania, March 18, 1856

Through a combination of chemical and mechanical means, this invention used a roller and crankshaft to pass previously prepared cloth through a box filled with a chosen paint created with special ingredients. Rollers then passed the cloth against blades to scrape off excess paint and then wound the cloth onto a second roller.

MACHINE FOR MAKING HORSESHOE NAILS, PATENT NO. 241,562
Erastus E. Pierce, New Brighton, Pennsylvania, May 17, 1871

The machine shaped a nail rod blank into the required shape. The nail rod was held by a rest and positioned so that a series of grinders could work a section of the blank until all four sides were ground to shape. The final stages would first straighten, then cut the completed nail.

BUILDING BLOCK, PATENT NO. 482,890
Carl Grunzweig, Ludwigshaffen-On-The-Rhine, Germany, September 20, 1892

To make these building blocks, water, clay and tar were mixed to a pulp and then cork chips were added. The mixture was poured into a mold for drying by heating. The heating and drying process evaporated the volatile components of the tar, leaving a tough, elastic product with corklike properties.

HORSESHOE MACHINE, PATENT NO. 230,180
Joseph H. Dorgan, Plattsburg, New York, July 20, 1880

A red hot blank cut from a bar of iron would be forced around a forming rod to create a "U" shape. The formed piece would then be held by top and bottom dies to be stamped with clearing holes and then bent to form the shoe.

PAPER BOX MACHINE, PATENT NO. 178,499
McClintock Young, Frederick, Maryland, June 6, 1874

The machine manufactures paper boxes from a roll of paper in one continuous connected movement. The paper was fed through a tube and forming roller to shape the bottom and sides of the box. Additional forming, pasting, drying, and printing steps completed the process.

CHAPTER 13

Paper & Printing

"The invention of the printing press was one of the most important events in human history."

— Ha-Joon Chang

The earliest evidence of printing dates back to China in the second century, where wooden blocks were used to transfer images of flowers onto silk fabric. In the 7th century, the Chinese printed the Buddhist text, the Diamond Sūtra, on paper, making it the first known (and surviving) printed book.

Practically all moveable type printing originates with Johannes Gutenberg, a goldsmith and stone cutter from Mainz, Germany. Gutenberg's wooden press mechanized the printing process. He developed the idea for moveable type printing around 1440 with his invention involving separate pieces of type for each letter, type which could be reused. Rather than carving entire words and phrases, Gutenberg carved the mirror image of each individual letter onto small blocks of wood. The blocks containing the letters could be easily moved, arranged and rearranged to form words. Gutenberg soon made use of his knowledge of metals, making his moveable type from an alloy of lead, tin, and antimony, known as type metal, printer's lead, or printer's metal. Casting type in metal was critical for producing durable type for high-quality, larger-run printed materials. Gutenberg was also responsible for introducing oil-based ink which was more permanent than the water-based inks previously used. In 1452, Gutenberg began printing the Gutenberg Bible for which he is also best known. Its high-quality and relatively low price proved the superiority of moveable type. Following his discovery, moveable type printing presses quickly spread throughout Europe, and eventually, around the world.

Until the 19th century, printers performed each step of the printing process by hand, as Gutenberg had done. As the technology improved, inventors dreamed up new techniques that accelerated the revolutionary import of printing. Two key ideas changed the design of the press itself — the use of steam power for running the machinery and the replacement of the flatbed for printing with the use of rotary cylinders. Both concepts were originally introduced by German printer Friedrich Koenig between 1802 and 1818. Koenig designed a steam press that he said was "much like a hand press connected to a steam engine." He patented his invention in 1810. The replacement of the hand-operated Gutenberg press by steam powered rotary presses allowed printing on a much greater scale.

In the late 18th and early 19th century, inventors modified the press further by fabricating parts of the press out of metal instead of wood. In 1800, Earl Stanhope of England, invented the Stanhope Press, the first press made completely out of cast iron.

The development of continuous rolls of paper, the steam-powered press, and a way to use iron instead of wood for press construction, all greatly increased the efficiency of the printing process. Richard M. Hoe first brought into practical use the rotary press in 1847. It was quickly adopted by nearly all of the daily newspapers in the U.S. as well as in England, Scotland, and Australia.

Until the late 19th and early 20th centuries, type was cast and composed by hand. The development of Linotype and Monotype machines improved the printing process further as they employed mechanical means for casting type, which was far more efficient than composing type by hand. (Linotype refers to a method of casting complete lines of type at a time while Monotype cast individual letters held in a matrix of letters.)

ROBERT HOE (1784-1833)

Robert Hoe was born in England in 1784. He studied carpentry before immigrating to the U.S. in 1803. With his brother-in-laws, Hoe formed a company in New York City in 1805 that specialized in the manufacture of wooden hand presses. As the company grew, they manufactured printing presses and machinery for the printing business and also various kinds of long and circular saws. Robert Hoe, Sr. was the first man in the U.S. to make saws of cast steel and the first in New York to power the machinery in his factory using steam. Robert Hoe tried to improve upon the hand-operated press, but it was his son, Richard who was to change the world of printing forever.

RICHARD MARCH HOE (1812-1886)

Richard M. Hoe was an American press maker who made many improvements to Friedrich Koenig's design, and in 1832, he invented the Single Small Cylinder Press. In this type of press, a sheet of paper is passed between a flat surface and a cylinder in which a curved plate is attached. The cylinder rolls over the flat surface and generates an impression on the paper. Cylinder presses were much faster than earlier platen or hand presses. In 1843, Richard Hoe invented the rotary press, which really revolutionized printing. The rotary press printed both sides of a sheet of paper at the same time. Nicknamed the "lightning press," it was powered by steam and used a revolving cylinder that could quickly turn out thousands of printed pages. The first rotary press could print up to 8,000 pages an hour. Larger rotary presses made printing large-run newspaper printing possible.

Not only did the Hoe Company manufacture printing presses, they also made improvements to existing machinery. After Robert Hoe Sr.'s death in 1833, his sons, Richard and Robert Jr. took over daily operations of the company, introducing many improvements for which they secured patents. In 1830, the Hoe Company was manufacturing a single small cylinder press. In 1837, the double small cylinder press, improved by Richard Hoe, was perfected and put into production. During that same period, Hoe designed and introduced the single large cylinder press—the first flatbed and cylinder press used in America. Hoe's rotary printing press was installed in the plant of the Philadelphia Public Ledger in 1847.

This rotary press was the predecessor of the web press (1871) eventually designed by Hoe and Stephen D. Tucker.

First installed in the plant of the *New York Tribune*, the web press simultaneously printed from a continuous roll of paper onto both sides of the sheet at a rate of 18,000 papers per hour. The Web Machine required only three people to run it. The Lightning Press, by contrast, printed only on one side and required eight feeders, one foreman, and two fly-boys (to lift the papers as they came off the press). Hoe also introduced the stop cylinder press (1853) and devised the triangular form folder (1881). The latter, in combination with the web press and the curved stereotype plate, is the foundation of the modern newspaper press used today.

Hoe's inventions were responsible for the company being one of the most successful manufacturers of printing equipment in North America. The 1876 Centennial Exhibition in Philadelphia featured a display of Hoe's web printing and folding machine, double cylinder, two revolution cylinder, stop cylinder, and lithographic printing machines and circular saws.

Today, 12 patent models by Richard Hoe are at the National Museum of American History in Washington, DC. Hoe received patents for over 30 inventions, several along with Stephen D. Tucker.

The Hoe Company was the only company to combine the production of all of the tools and machinery of printing, lithographing, stereotyping, and electrotyping. Everything having to do with the process of printing was manufactured by them. In the 1800s, electrotyping was a standard method of producing plates for letterpress printing. Stereotyping was a process where a mold was cast from an original and then used in the printing process instead of the original. (Figures 1 & 2)

MACHINE FOR PRINTING RAILROAD AND OTHER TICKETS, PATENT NO. 23,172

Richard M. Hoe, New York, New York, March 8, 1859

A new printing press for printing and numbering railroad and other tickets. The tickets were printed on a strip or roll of paper by types, or stereotype plates, or solid raised or sunken letters. A revolving cylinder numbered the tickets with successive numbers by a registering apparatus, and than cut them from the roll and deposited them in a receptacle in numerical order. A separate inking apparatus allowed the numbers to be printed in a different color.

Figure 1: R. Hoe and Co. Factory

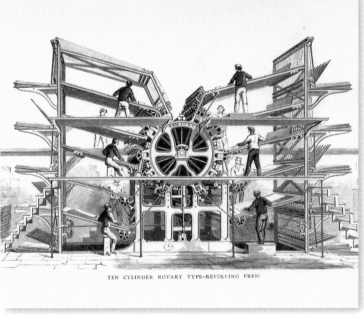

Figure 2: R. Hoe and Co. Ten Cylinder Rotary Press

STEPHEN DAVIS TUCKER (1818-1902)

Stephen Tucker was associated with R. Hoe & Company for almost 60 years. Tucker entered the business as an apprentice in 1834 and rose in the ranks to become a partner in 1860.

In 1877 alone, over a dozen patents were granted to Tucker relating to web printing and delivering mechanisms associated with the process. Tucker held individual patents, as did Richard Hoe, but they also held patents together.

PASTING MECHANISMS FOR PAPER FOLDING MACHINE, PATENT NO. 212,872
Stephen D. Tucker, New York, New York, March 4, 1879

In this device, a rotating pasting blade applied a transverse line of paste to the sheet of paper as the paper passed over the supporting cylinder. When the pasting blade came in contact with the supporting cylinder, it removed the line of paste so it did not stick to the inner surface of the next sheet of paper.

MACHINE FOR FOLDING PAPER, PATENT NO. 188,987
Stephen D. Tucker, New York, New York, March 27, 1876

In this machine, used with a web printing press, a rotating sheet carrier was equipped with a vacuum sheet-holding device. It first folded the sheet of paper, and then severed it from the web or roll in one continuous action and could print on both sides.

FRANCIS WOLLE (1817-1893)

Francis Wolle patented his paper bag machine in 1852 in the U.S. and later in France and England. Wolle was a botanist, clergyman, and teacher who was born in Jacobsburg, PA in 1817. Wolle was educated in the Moravian parochial school in Bethlehem and then became a clerk in his father's store. In 1869, Wolle and his brother and other paper bag makers together founded Union Paper Bag Machine Company. The Union Paper Bag Machine Company is still in business today in Savannah, GA and is owned by International Paper.

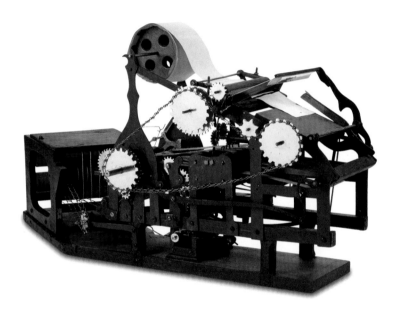

MACHINE FOR MAKING PAPER BAGS, PATENT NO. 12,982
Francis Wolle, Bethlehem, Pennsylvania, May 29, 1855

This machine created a complete paper bag, from cutting, to folding, pasting, and printing, and finally to drying. The drying chamber's "peculiar oblique position" left the bags' pasted surfaces exposed to heat so they dried properly.

PAPER

Paper was invented in ancient China and spread slowly to the west via the Silk Road. The Chinese developed "rag" paper, a cheap cloth-scrap and plant fiber substitute for previous types of paper made of bark and bamboo strips or even silk.

In Europe, white paper was the most desirable type. Poorer grades of paper were made of old, discarded materials and were between a light coffee to a light grey in color. Bleaching was unknown in the early 19th century, so papermakers had to depend on using only a better quality of pulp fibers to get a brighter sheet. The best fabric used for paper was a linen of the whitest kind. The cotton and linen of the period were woven by hand and were free of chemicals and bleaching.

Paper making became mechanized with the use of waterpower. The rapid expansion of European paper production was due to the invention of the printing press and the beginning of the printing revolution.

The first paper mill in the U.S. was established in 1690 in Philadelphia. It was a recycled paper mill, using rags to manufacture its product. Thomas Gilpin is credited with manufacturing the first machine-made paper in the U.S., in 1817. It was not until the introduction of wood pulp in 1843 that paper production became independent of recycled materials obtained from ragpickers.

In the 1830s and 40s, Charles Fenerty and Friedrich

Gottlob Keller experimented with wood pulp using the same techniques already employed in paper making. Instead of pulping rags, they pulped wood. The pair invented a machine which extracted the fibers from the wood. Fenerty also added bleaching to the process so the resulting paper was white. The innovations of these two launched a new era in paper making. By the end of the 19th century, almost all printers in the western world were using wood-based paper.

Steam-driven rotary presses and wood-based paper marked a great turning point for the printing industry and the media that employed them. Suddenly, paper and printing was cheaper, easier, quicker and more accessible to a much greater segment of the population. (Figure 3)

Figure 3: Paper Cylinder Machine

PAPER CUTTING MACHINE, PATENT NO. 198,519
John G. Morgan, Appleton, Wisconsin, December 11, 1876

The vertical movement of this machine's knife resulted in a cleaner cut and a smoother edge on the paper than did cutting laterally.

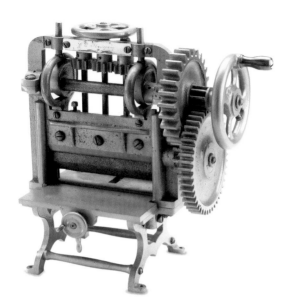

PAPER FOLDING MACHINE, PATENT NO. 35,738
John North, Middletown, Connecticut, June 24, 1852

Prior to this invention, paper folding machines used tapes to carry the paper through the machine. North's improved machine eliminated the "annoyance of the endless tapes."

MACHINE FOR COLORING PAPER, PATENT NO. 85,426
Charles K. Brown, Troy, New York, December 29, 1868

Prior to this invention, colored paper was applied by hand or by a device that brushed it on only one side of the paper. Brown's improvement used rollers to pass a roll of paper through a reservoir of coloring matter to both sides of the paper at the same time.

MACHINE FOR MAKING PLAITED PAPER CAPS, &c., PATENT NO. 221,314
William F. Hunt, London, England, November 4, 1879

This invention improved machines for manufacturing plaited paper caps, capsules for bottles, paper cups, cases, holders, or other receptacles for holding ices, custards, soufflés, ramequins, cheese, fruit, and other refreshments.

PAPER FOLDING MACHINE, PATENT NO. 225,506
Cyrus Chambers Jr., William Mendham, Philadelphia, Pennsylvania, March 16, 1880

This machine improved on the inventor's earlier, Patent No. 114,489. The invention folded damp limp paper very rapidly. The folds are "imparted to the sheet by rotating folding blades and by grippers attached to the folding rollers, all the movements of the machine being rotary and continuous, thus insuring the highest practicable rate of speed in the performance of its work."

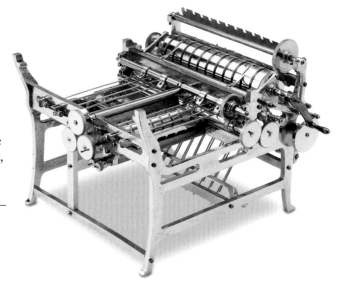

RECIPROCATING PRINTING PRESS, PATENT NO. 147,257
John F. Hallenback, Port Chester, New York, February 10, 1873

Hallenback's printing press placed the type in a frame placed on the top of a large stationary cylinder. Around the cylinder were a series of ink rollers. The interior of the cylinder contained an ink reservoir, which distributed ink evenly to the inking rollers. An adjustable padded platen moved the frame up and down during the printing process. The printed paper was carried between rollers, cut into the required lengths and deposited into a box in front of the press.

TICKET PRINTING AND RECORDING MACHINE, PATENT NO. 209,827
John Moss, John H. Smith & George J. Hill, Buffalo, New York,
November 12, 1878, Assignors to The Cash Recording Machine Company,
New York, New York

The machine registered the value of each transaction on a paper strip kept inside the machine, and stamped that value on each customer account bill. "Intended to be a check on employees, and a prevention of dishonesty by clerks."

MACHINE FOR MAKING PAPER COLLARS, PATENT NO. 72,705
Oscar F. Washburn, Bridgewater, Vermont , December 24, 1867

The detachable paper or fabric collar was invented by Hannah Montague, a housewife in Troy, NY, who was tired of trying to remove the "ring around the collar" from her husband's shirts. A separate collar for a shirt was both more efficient for laundering and also more economical because a soiled or frayed collar could be replaced without having to buy a whole new shirt. Washburn's machine could cut, emboss, button hole, and crease such a collar in a single operation.

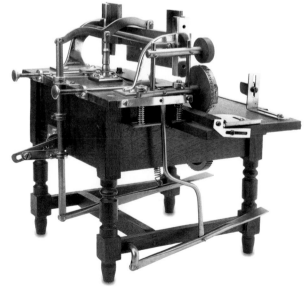

MACHINE FOR CUTTING AND PRINTING INDEXES, PATENT NO. 168,468
Henry H. Edwards, Grand Rapids, Michigan, October 5, 1875

An improved machine for indexing books. The machine's main table supported the entire cutting, printing, and operating mechanism, and the book supporting and feeding table could move to the right and left to cut and print the leaves.

ROTARY PAPER CUTTER, PATENT NO. 164,920
Agur Judson, Newark, New Jersey, June 29, 1875

This machine cut strips of paper for telegraphs and other uses, using a series of disk cutters fixed on a revolving shaft that was applied to a second shaft of disk cutters. The cutters act as partial feeders to prevent tearing of the strips.

SURFACES FOR PRINTING ON METAL, PATENT NO. 220,549
John M. Ronemous, Baltimore, Maryland, October 14, 1879

A method for ornamenting tinware, using a thin sheet of rubber cemented to a smoothly dressed, level-surfaced board. The rubber sheet was engraved with the desired figure or ornament, and this printing plate was pressed or rolled over the tin to transfer the print impression upon it.

PRINTING PRESS, PATENT NO. 197,021
Edward Dummer, Boston, Massachusetts, November 13, 1877

An oscillating printing press that carried inking rolls over the form using "bands or chains passing about pulleys rotating continuously in one direction." The inking rolls passed over the form only once for each impression.

CHAPTER 14

Textiles

"Clothes and manners do not make the man; but, when he is made, they greatly improve his appearance."

— Henry Ward Beecher

The textile industry was among the first of America's industries to be mechanized, with textile workers becoming America's first generation of factory workers. It was the textile industry, in the late 18th and first half of the 19th centuries, that fired the first shots across the bow of America's Industrial Revolution.

That revolution coincided with America's other revolution, its war of independence from England. As the country fought for and won that independence, Americans clearly saw the colonial nature of their economy and realized that new mechanized technologies offered tremendous economic advantages that had been unavailable to them under the yoke of colonialism.

The Industrial Revolution would take the U.S. from a local to a national economy, and then fully onto the international stage, replacing man and horsepower with machines, and relocating thousands of its workers from the country to the city. And the first industry to head up this march of progress was textiles.

THE PROCESS OF MAKING CLOTH

Making a piece of cloth involves a number of steps. The first is the harvesting and cleaning of the fiber or wool. From there, the wool is carded (combed) and spun into threads. Next, those threads are woven into fabric, and finally, the cloth is designed and sewn into its intended form for final use.

Before the mechanization of textile production, spinning was an essential household task. The simple spindle was the earliest spinning tool, which was gradually replaced by the spinning wheel. Almost every household, from Colonial times to the mid-1850s, owned a spinning wheel. Originally, all clothes were made in the home. Cotton and woolen yarns were spun by hand, woven or knit into coarse fabrics, and then made into various articles of clothing. There were early efforts to mechanize spinning, or at least increase its productivity. Leonardo Da Vinci left a drawing of a spindle that traveled back and forth to automatically distribute the yarn evenly on a bobbin, and Giovanni Branca left plans for a spinning wheel powered by a waterwheel.

The American colonies were dependent on England for everything but homespuns (fabrics for clothes woven in the home). During the early 18th century, in its desire to dominate the textile industry, Great Britain passed laws forbidding the export of English textile machinery, drawings of machinery, or written specifications that would allow the machines to be constructed abroad.

These laws made it impossible for colonists to establish their own textile industry. It wasn't until 1845 that spinning and weaving machinery were finally permitted to be exported from the United Kingdom. During all these years, not a single complete textile machine was successfully smuggled into the U.S.

By the mid-1850s, through mechanization, spinning began to move outside of the home and into factories. This process accelerated in the late 18th century with the introduction of steam power.

Sir Richard Arkwright of England is an important figure in the history of textile machinery. Arkwright's machine improved upon the existing equipment of the time. His device did not stop to wind lengths of yarn, as did the spinning wheel, but rather performed the entire process of spinning. Arkwright also introduced the factory system (of steam-powered manufacture, division of labor, interchangeability of parts) into the textile industry.

In the late 1850s, Isaac Singer built the first commercially successful sewing machine with a needle that moved up and down rather than side to side, powered by a foot treadle. Besides designing and manufacturing the machine, Singer was also responsible for the first large-scale consumer appliance business, with such innovations as consumer advertising, selling machines on an installment plan, and offering a service contract. Singer received 200 patents on his sewing machine.

The changes in textile production associated with the Industrial Revolution were mainly related to spinning and weaving. John Kay's flying shuttle of 1733 reduced human effort in weaving on a loom. James Hargreave's spinning jenny of 1770 allowed a single operator to spin several yarns at the same time. Using spinning frames with multiple spindles and powered by water came next. Power looms appeared later, first used with cotton, then with wool. The introduction of steam power into textile mills freed the industry from the limitations of water power and allowed mills to be built anywhere.

The Boston Manufacturing Company, (also called the Waltham Company) founded in Waltham, MA in 1813 by Francis Cabot Lowell, was the the first textile mill in the world in which the whole process of cotton manufacturing, from spinning to weaving, was completely non-human powered. The power loom was the first step toward creating an integrated textile mill. The new loom required yarn stronger than that used in hand looms, and therefore, new types of

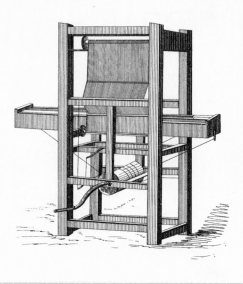

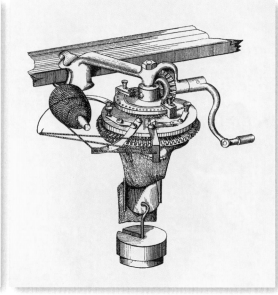

Figure 1: Weaving Loom c. 1875   *Figure 2: Circular Knitting Machine c. 1880*

spinning and sizing machines were suddenly needed. The Boston Manufacturing Company made so many improvements in milling machinery that much of the firm's early profits came from selling patent rights for its inventions. But the biggest invention to come out of Waltham was the integrated factory which mechanized all of the processes of textile production under one roof. (Figure 1)

Francis Cabot Lowell's Machine Shop was one of the foremost textile machine manufacturers in the U.S. They not only made almost all of the machinery used at the many mills of Lowell, but also sold textile equipment to mills throughout the country. The same tools used to produce textile machines could also be used to make other machines. In the "Waltham-Lowell System," for the first time, both spinning and weaving occurred on site and mill workers resided in collective company housing under strict supervision.

America's first planned factory town was Lowell, MA. The mills of Lowell were on a new scale, much larger and more industrial than anything seen before. Lowell became known as the "Manchester of America," a new industrial city, built from scratch by a group called the Boston Associates. These same men had earlier invested in the Boston Manufacturing Company.

The mills of Lowell set the bar for the textile industry of northern New England. By 1860, the textile industry was an important part of the economy in all parts of the country. Its effects were felt in the daily lives of almost every American. Consumers no longer had to spin and weave their own clothing and mass-produced textiles meant cheaper clothing for all. With its large work force, many machines, and heavy use of water and steam power, the textile industry became iconic of American's early Industrial Revolution. (Figure 2)

SAMUEL SLATER (1768-1835)

Samuel Slater began his career in one of Richard Arkwright's English factories. A pioneer of cotton manufacture in the U.S., Slater was born in Belper, Derbyshire, England. Apprenticed to Jedediah Strutt, a partner of Richard Arkwright in the development of cotton textile machinery, Slater learned all aspects of the textile trade. He memorized the details of the machinery that the company produced, and then, encouraged by the bounties being paid for the introduction of cotton textile machinery into the U. S., he emigrated to the U.S. in 1789. Slater reproduced Arkwright's cotton machinery from memory, building his first plant in Pawtucket, RI. He made the machinery with his own hands, working entirely off of Arkwright's plan. His machinery was the first of its kind ever operated in America, and inaugurated the American cotton industry. Slater succeeded where others failed because he had worked for one of the largest textile manufacturers in England as a mill overseer. Besides the technical skills (and machine plans) he learned there, he also gained managerial skills that would prove invaluable in the U. S. The success of Slater's RI mill led to the expansion of the textile industry throughout New England.

Slater continued to modify and adapt his machines, but the development of a larger cotton industry became a reality with the invention of the cotton gin. Mechanical improvements had brought cotton manufacturing to the point where the need for raw material was greater than supply. The price of cotton was very high, and the chief expense and major difficulty was cleaning it of the dirt, leaves, and seeds carried with it from the fields in its fibers. It was thought that cotton could not be raised in larger quantities because of the high cost of cleaning it and preparing it for the spinner. Eli Whitney's invention, the cotton gin (see Chapter 3: Alcohol, Tobacco, and Firearms (ATF) for more), which could clean harvested cotton with great speed, created a new demand for cotton fabrics.

In 1804, Oliver Evans, of Philadelphia, developed a high pressure steam engine. It was adaptable to a great variety of industrial purposes including powering textile factories.

BRAIDING MACHINE

Thomas Walford was the father of the braiding machine, used to produce worsted (yarn spun from combed long-staple wool) dress braids. Walford, of Manchester, England received one of the first British patents for a braid machine in 1748. The significance of Walford's invention was the carrier and the ratchet bobbin which released yarn on demand. Almost all carriers in use today use a ratchet release similar to Walford's.

Braiding machines were first manufactured in the U.S. in 1861. Most of the braid factories were located in New England. Prior to this, smaller braiders for the covering of whips and hoop skirt wire, and the production of shoe and corset lacings, as well as other small items, had been made in New England. But broader braids were all imported and no machines existed in the U. S. In 1861, Gilman K. Winchester of the Rhode Island Braiding Machine

Company of Providence developed a braiding machine considered far superior to the English model, which was cumbersome and expensive in comparison.

Braiding turns fibers into more useful forms. Smaller fiber strands are braided in order to produce larger, stronger ones, like rope. On a standard braiding machine, the supply lines are at a constant angle and a constant tension to insure that the output is uniform. The braiding machine emulates a May Pole forming a downward braid on the pole by alternating strands of yarn.

MAKING BONNET BINDING, PATENT NO. 46,163
Jefford L. Weaver, Orange, Massachusetts, January 31, 1865

It was customary to weave bonnet braids of palm leaf in short lengths, and then arrange them together as warps in a weaving machine, then a weft of palm was woven into them resulting in a ribbon approximately two feet in length. Previous methods had only resulted in strips two feet or less in length. This machine made braids of any length.

SHUTTLE MOVEMENT FOR LOOMS, PATENT NO. 118,114
John Detweiler, West Liberty, Ohio, August 15, 1871

Looms of the ordinary type have a "shuttle" that holds the filling yarn that is usually passed back and forth on a rail or "sleigh board" by hand as the warp yarn is reciprocated up and down to produce a weave. This manual practice was know as "throwing the shuttle." This machine was designed to pass the shuttle back and forth automatically through various mechanical movements as the machine was put into motion. Triggers would release and reset the spring works as the shuttle passed over them. The operation would stop if the shuttle failed to trigger the spring due to entanglement or other faults.

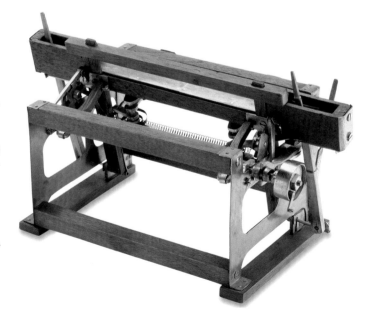

MACHINE FOR CLEANING WOOL FROM BURS & OTHER FOREIGN MATTER, PATENT NO. 925

Theodore Ely, , Poughkeepsie, New York, September 17, 1838

This machine removed burs and "foreign matter" usually found in imported wool from South America and elsewhere. The operator fed the wool into a set of rollers, and it passed over a blade of sheet steel that scraped the foreign matter out of the wool. The back of the machine contained a fan that would blow the cleaned wool out and away from the machine.

KNITTING MACHINE, PATENT NO. 75,353

Henry Bogel, Watertown, Wisconsin, March 10, 1868

This knitting machine made plain knit fabric of any number of stitches, with two independent sets of needles, each working independently of each other. This arrangement enabled the operator the ability to knit one or two pieces of fabric at the same time. Simple in construction, it worked without producing very much noise and was easy to take apart for maintenance, cleaning, and the task of both removing and replacing needles.

MACHINE FOR MAKING ROPE, PATENT NO. 9,414

Hezekiah T. Jennings, Charles S. Collier, Bethany, New York, Thomas P. How, Buffalo, New York, November 16, 1852

This machine kept tension on the formed rope to prevent entanglements throughout its manufacture. Starting from bobbins that fed twisted individual strands towards a cone shaped laying block, the forming strands were twisted to form the rope and then would be pulled over an enclosed receiving reel by which the rope was drawn. The speed of the receiving wheel was regulated by the amount of tension on the rope.

CIRCULAR BRAIDING MACHINE, PATENT NO. 81,038
William Tunstill, Patterson, New Jersey, August 11, 1863

A typical "stop action" mechanism was adapted into this braiding machine. Each spool on the bed revolved around the center strand, creating the braid. If a thread broke—which was common—the spool and attached weight dropped to the level of projecting tabs positioned inside the ring. As the dropped spool made contact with the tab, it partially rotated a ring which activated a lever that disengaged the main drive gear to stop the machine.

TAILOR'S TABLE, PATENT NO. 207,575
Albin Warth, Stapleton, New York, August 27, 1878

This table, which was directly connected to a sidecar with a seat for the operator, had a package carrier that supported and fed out a roll of fabric. The car and package traveled together in unison as the fabric was laid out along the table, usually forty to sixty feet long. The operator controlled the movement using foot pedals.

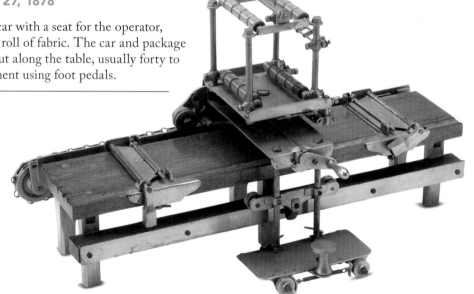

MACHINE FOR BRAIDING MANILA, PATENT NO. 1,566
Daniel, Jesse & Elisha Fitzgerald, New York, New York, April 24, 1840

This machine, made to braid Manila or Sisal hemp, consisted of causing spools of fiber attached to "carriers" to traverse a serpentine path two thirds around a "pot" and back, interlacing the fibers to produce a braid.

COTTON PRESS, PATENT NO. 187,814
Stillman A. Clemens, Chicago, Illinois, February 27, 1877

A cotton gin was directly connected to this press which could then bale the ginned cotton by a continuous automatic action. The cotton, as it came from the gin, was formed into a continuous sheet, which was pressed and laid under pressure in folds doubled up one upon the other. Screws slowly released by the feed mechanism allowed the bed to descend to admit more cotton. The bale was compressed under the machine in the baling bed, to be tied or hooped.

KNITTING MACHINE, PATENT NO. 9,637
John Maxwell, Galesville, New York, March, 29, 1853

Specific changes related to the movement mechanism used in typical looms in use at this time were incorporated in this new knitting machine. A typical loom of this type, when bringing a cloth over a new made stitch, would produce two violent motions due to cam design and would require levers and arms of great strength to endure the work, limiting the speed of the machine. Unlike the "violent motions" of other knitting machines, this new loom's redesigned cam works caused much less stress on the knitting machinery. Innovative changes to the mechanics also included the use of springs to regulate the pressure in key areas of the operation of the loom.

IMITATION FEATHER, PATENT NO. 282,661
Henrietta S. Orttlopp, John C. Kloberg, New York, New York,
August 7, 1883

This machine could produce an imitation ostrich feather or similar
feathers from an ordinary worsted wool yarn. Very specific looping and
stitching patterns were detailed in the patent paper description. Wood,
whale bone, or metal could form the quill of the feather.

FABRIC FOR WALL DECORATION, PATENT NO. 345,191
Thomas Strahan, Chelsea, Massachusetts, July 6, 1886

Imitation silk or satin produced a cheap and beautiful damask fabric
wall hanging. The cotton, linen, or jute fabric would first be dyed
or tinted then calendared through rollers to produce a smooth and
lustrous finish. A design could then be printed, stamped, or stenciled
upon the calendared surface.

CHAPTER 15

Marine & Navigation

"Twenty years from now you will be more disappointed by the things you didn't do than by the ones you did do. So throw off the bowlines. Sail away from the safe harbor. Catch the trade winds in your sails. Explore. Dream. Discover."

— H. Jackson Brown, Jr.

As a developing U.S. expanded, waterways became the nation's early highways. Sea, river, and canal navigation all developed hand-in-hand with a nation's growing economy. Waterways provided cheap transportation for freight and people. Early American commerce grew up along waterways before expanding inland. While rivers could serve as highways, they also presented dangers, with rapids, falls, and other water-borne perils. Man-made canals were developed and built as a way of overcoming these natural obstacles. Water travel could also be slow and depended on winds, river currents, and manpower. This changed with the invention and introduction of the steam engine and steam-powered boats.

The first canals built in the U.S. were no more than twenty miles long. Canal boats were powered by horses and mules drawn on pathways along the canals. East coast ports needed connections with the Great Lakes for business and trade. The Erie Canal, proposed in 1807, was a bold idea, linking Buffalo on Lake Erie, with Albany on the Hudson River. Since the Federal government refused to fund it, Dewitt Clinton, governor of New York, proposed that New York State should undertake the construction of the 363 mile long canal itself, which ended up costing over seven million dollars (in 19th century dollars). Detractors nicknamed it "Clinton's Big Ditch." The Erie Canal was, however, a great success, opening in 1825, and leading to the growth of the cities of Buffalo, Cleveland, Detroit, and Chicago.

In the 19th and 20th centuries, most steam boats were actually steam-powered paddlewheel boats, with wooden hulls and one or two wooden paddlewheels. The paddlewheel had a tubular center with spokes radiating from it like a bicycle wheel. Planks stretched across these spokes served as water paddles, and could be placed on either the side of the boat or at the rear. Sternwheelers had a single paddle at the rear, while sidewheelers had paddles on both sides. The paddle wheelers were powered first by wood, later by coal. Using steam to reliably power these vessels made inland waterways an even more valuable resource to a growing, industrializing America.

The biggest danger in steam engine technology was the threat of boiler explosions. If boilers were not carefully watched over and maintained, pressure could build up rapidly and cause them to explode.

In the 1800s, crops, manufactured goods, machinery, and raw materials were all shipped over our nation's rivers and canals. The main means for transporting the important crops of cotton and tobacco were mainly over the riverways of the time. Different types of boats served different functions in this water-borne transportation system. Towboats moved barges by pushing them up and down riverways. Ferries carried people back and forth across the rivers. Snag boats cleared the river of dangers (a snag being sunken trees, stumps, or boat wrecks that could cause damage to a ship). Packet boats carried goods, mail, and people. Steamboats called fuelers met other steamboats along the river and re-supplied them with wood, coal, or oil. The most famous type of steamboat was the showboat. Showboats were the floating palaces of the age. They had theaters on them, galleries, ballrooms, and saloons. These grand, majestic vessels traveled up and down the river, bringing theater, music, and other types of entertainment to towns and cities along its shores.

Although steamboats ruled trade and travel throughout the 1800s and early 1900s, they were eventually replaced by newer forms of transportation, including ground-based transportation like expanding roadways and rail lines. Steamboats began experiencing competition from railroads as early as the 1830s.

The year 1853 marked a high point in wooden clipper ship building. The railroad and communications and transactions over telegraph contributed to this decline. The sailing ship soon yielded to the steamer for both passenger service and freight transport. Metal, and eventually steel, began to be used in ship construction, replacing wood.

Travel over water continued to become safer throughout the 1800s, but storms, fires, and coastlines still threatened seafarers. Collisions were frequent, as were steam explosions. And there was always the possibility of a shipwreck. Safety at sea was of great concern in the 1800s and the inventions of the time reflect that. Inventors foresaw the potential for disaster as the seas grew busier with human activity. The U.S. Patent Office, between 1790 and 1873, granted 163 patents for a variety of life-preserving boats, rafts, clothing, and other gear. The items included life-preserving bedsteads, berths, buckets, bucket rafts, buoys, capes, chairs, stools, dressers, doors, garments, hammocks, mattresses, even a "life-preserving hat"! Few of these inventions found widespread success.

ROBERT FULTON (1765-1815)

Robert Fulton was known as the "father of steam navigation," since he succeeded in turning the steamboat into a commercial success. Fulton was not the first person to build a steamboat. An Englishman, Jonathan Hull, had patented a design for a steamboat in 1737 and Americans James Rumsey, John Stevens, and James Fitch had all plied American rivers with steam before Fulton launched the Clermont in 1807. But those boats had been expensive to build and to operate. The Clermont was the first successful steamboat. Fulton was successful because he was backed by one of the richest men in America, Robert Livingston. Livingston held two monopolies on steam navigation. He started on the Hudson River, where land transport had a harder time competing. Fulton took the ideas of earlier inventors and improved on them, turning the steamboat into a commercial success.

Fulton was born in Lancaster County, PA. He eventually moved to London to turn his lifelong interest in science and engineering into a career. He had previously invented a submarine, a marble cutting machine, and several types of bridges. Fulton was especially interested in applying steam engines to navigation and in the design of canal systems. He moved to France to work on canals, and it was there that he met Robert Livingston. Livingston was a lawyer from New York who had served in the Continental Congress and on the committee that drafted the Declaration of Independence. At the time of his meeting Fulton, Livingston was the U.S. Ambassador to France and one of the richest men in America. The two men partnered to build a steamboat capable of navigating the Hudson, and revolutionized navigation in the U.S. in the process. Fulton established steamboat commerce on the Hudson and on the Mississippi, and was planning a similar venture on the Chesapeake, when he died in 1815.

CAPTAIN JOHN DOMINIS (1796-1846)

John Dominis (Figure 1) was a ship captain who arrived in New England from Trieste in 1819. As the captain of a trading vessel out of Boston Harbor, he did a great amount of business in China, and also in the fur and salmon trades on the Pacific coast. Dominis moved with his wife and young son, John Owen, to Honolulu, in 1837. In 1842, he received a patent on an instrument for measuring canvas for ship sails. He signed the application "John Dominis of the Sandwich Islands." After having made what in those times was a fortune ($96,000), he built Washington Place, then referred to as "the grandest mansion in all Honolulu." In 1846, he set off for China, in search of exotic oriental furnishings for his new home, but he was never heard from again. He was presumably lost at sea. Dominis never got a chance to live in the mansion he had spent four years building, but his wife Mary and their young son John moved in, taking in boarders to pay her bills. Until she died in 1889, Mary Dominis waited and watched from the windows of Washington Place, hoping to see her husband return. Washington Place has, since 1922, served as the official residence of the Governor of Hawaii. In 2007, it was designated as a National Historic Landmark.(See Figure 1)

John Dominis Jr. grew up to marry the last Queen of the Sandwich Islands (Hawaii), Queen Liliuokalani, and to become the Governor of Oahu and Maui.

Figure 1: Portrait of Captain John Dominis

INSTRUMENT FOR MEASURING SAILS, PATENT NO. 2,790
John Dominis, Sandwich Islands, Pacific Ocean,
September 30, 1842

Sails were generally constructed by sewing together strips of cloth or canvas, and much material was wasted in the process. This device, along with the inventor's mathematical scales and tables, permitted sailmakers to size and cut a roll of canvas more accurately. The scales and tables led to better layouts, so the needed shapes of material could be cut with minimal waste.

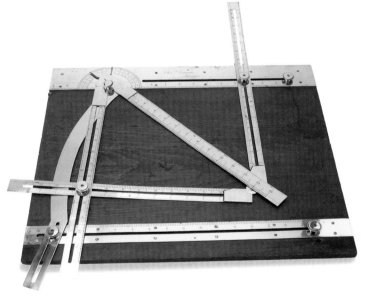

HENRIETTA VANSITTART (1833-1883)

No woman epitomizes the idea of the new woman inventor more than Henrietta Vansittart, of Richmond, England. Henrietta is known as England's first woman engineer, and her improvements in propeller design is probably the most important nautical invention by a woman of the 19th century. Her ship propulsion design is especially noteworthy because it is such an unusual area of invention for a woman of her time. In 1838, Henrietta's father, James Lowe, had patented his idea of a submerged stern-mounted propeller to replace the partially submerged side-mounted paddlewheels then in use. Although the Lowe propeller was fitted on many British warships, James Lowe never saw any financial benefits from his invention. Infringement battles prevented him from any monetary reward, and as a result, he died penniless in 1866.

Figure 2: Portrait of Henrietta Vansittart

Following her father's death, Vansittart carried on his pioneering work on the development of the screw propeller for steamships. The experience of her father appears to have been a driving force in Henrietta's pursuit of her invention. Determined to win the recognition for her father that had eluded him during his lifetime, she began to experiment, and in 1868, she was awarded British patent #2877. Her propeller was later used in many warships and liners, including the Scandinavian and the Lusitania. She took out both English and American patents for her invention.

In May 1869, Vansittart received a U.S. patent for her improved method of construction for screw propellers. Her propeller allowed ships to go faster, with less fuel, less vibration, and better steering in reverse. (Figure 2)

METHOD OF CONSTRUCTION FOR SCREW PROPELLERS, PATENT NO. 89,712
Henrietta Vansittart, Richmond England, May 4, 1869

This invention minimizes the power required to drive a ship's propellers, preventing "churning" or useless stirring of the water near the center of motion. The invention defines a method of determining the appropriate blade form and curvature using factors such as the length of stroke of the engines, the number of RPMs, the horsepower supplied, and the intended velocity of the vessel through the water.

DAVID BROWN (1793-1850)

One of the most influential clipper ship building firms in New York was that of Jacob Bell and David Brown. David Brown was one of the best ship designers of his day, with the firm designing and building both sail and steam vessels. The partnership of Brown and Bell was a long and profitable one, lasting until Brown's death in 1850. Brown and Bell built two early clipper ships together, the Houqua and the Samuel Russell.

The Houqua, built in 1844, was 583 tons, built for A.A. Low & Brother of New York as a warship to be sold to the

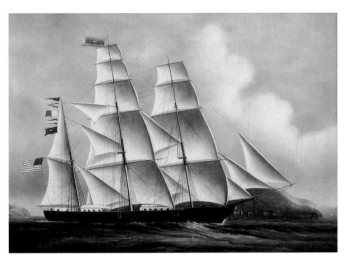

Figure 3: The Clipper Ship "Houqua" c. 1850

Chinese government. After it was determined that it was too small for its intended purpose, it was used almost entirely in Chinese trade. The Houqua was notable for having an innovative hull design. She was named in honor of the Canton Hong merchant Houqua, with whom the Low brothers had traded with in China for many years. (Figure 3)

The Samuel Russell (940 tons) was built in 1847 for A.A. Low & Brother and was named for the merchant founder of the house of Russell & Company, in China with whom the Low brothers began their business career.

SHIP BUILDING, PATENT NO. 5,344
David Brown, New York, New York, October 30, 1847

David Brown, a naval architect and ship builder, improved the construction and combinations of materials and parts used to build ships of all sizes or classes for any purpose. The improvements increased strength and decreased hull weight. The improvements consisted principally of methods for applying iron conjointly with wood to the frame timbers and beams of the ship being built.

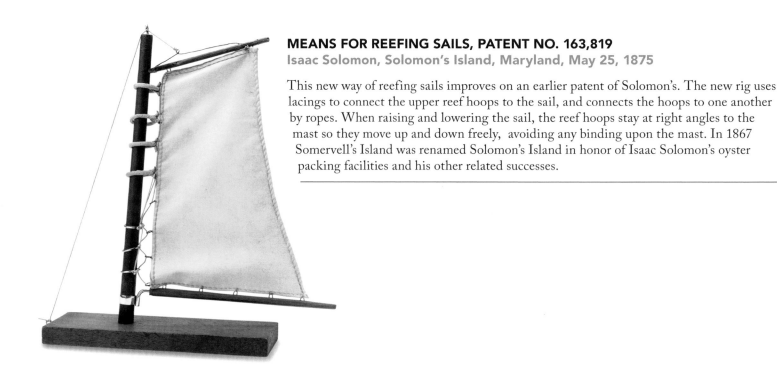

MEANS FOR REEFING SAILS, PATENT NO. 163,819
Isaac Solomon, Solomon's Island, Maryland, May 25, 1875

This new way of reefing sails improves on an earlier patent of Solomon's. The new rig uses lacings to connect the upper reef hoops to the sail, and connects the hoops to one another by ropes. When raising and lowering the sail, the reef hoops stay at right angles to the mast so they move up and down freely, avoiding any binding upon the mast. In 1867 Somervell's Island was renamed Solomon's Island in honor of Isaac Solomon's oyster packing facilities and his other related successes.

LIFE PRESERVING STATE ROOM FOR NAVIGABLE VESSELS, PATENT NO. 20,426
Henry Hallock, Brookhaven, New York, June 1, 1858

This Jules Verne-like invention consists of an airtight stateroom for steamboats or other vessels. In the event of a ship sinking, the stateroom is designed to automatically detach itself and float off the deck of the vessel. Once detached, the boat drift could be controlled by the occupants, store food and water, and provide light and ventilation until they are rescued.

MODE OF COMPENSATING THE LOCAL ATTRACTION OF THE MAGNETIC NEEDLE ON SHIPS, PATENT NO. 16,845

Calvin Kline, New York, New York, March 17, 1857

Compasses are prone to "local attraction" caused by the natural magnetism of the earth, which affects the accuracy of the needle. This invention compensated for local attraction by surrounding the compass needle with an insulated ring shield of iron and steel. The compass could still travel as freely and settle as quickly as an ordinary compass.

SEA DRAG, PATENT NO. 94,141

Jacob Edson, Boston, Massachusetts, August 31, 1869

This sea drag or sea anchor provided a cheap, strong, reliable, and compact drag for governing a vessel during a storm. The drag's four wings or arms spread open by means of rods and levers when it was thrown in the sea. The line attached to the drag automatically closed the four wings as it was hauled back aboard, minimizing resistance. For storage, the drag collapsed into a boxlike shape.

FEATHERING PADDLE WHEEL, PATENT NO. 216,208

William T. Merritt, Poughkeepsie, New York, June 3, 1879

A feathering paddle wheel adjusts the angle of the paddle while in motion. The paddle enters the water perpendicularly, then positions itself to propel, then exits perpendicularly. Feathering greatly reduces wasted power and water disturbance that can damage riverbanks and endanger nearby smaller water craft.

This invention's improvements center on the mechanics of the paddles, better controlling their orientation using sliding stops and cams.

COMBINED TABLE AND LIFE PRESERVER, PATENT NO. 208,473
Henry M. Green, Camden, New Jersey, October 1, 1878

This life preserver was also a watertight table with waterproof lockers for valuables. In the event of a shipwreck, passengers could remove the table's top leaves and throw the table overboard. The buoyant table body would become a life preserver, floating with its legs facing upwards. Lashings were provided for people to hold themselves to the preserver.

LIFE PRESERVING BERTH FOR STEAM & OTHER VESSELS, PATENT NO. 18,090
Eldridge Foster, Hartford, Connecticut, September 1, 1857

This movable and adjustable sleeping berth had inflated and elastic keels so it could double as a life preserver. In the event of a shipwreck, passengers were to fold the keels out of the bottom, thus allowing air to enter through valves, filling an elastic waterproof cavity. Bench seats folded out to accommodate two or three persons. According to the inventor, "Each person when using this berth will feel confident that he is sleeping upon that which will perhaps in case of accident be the means of his safety."

PROPULSION OF VESSELS, PATENT NO. 84,602
Albert F. Yardell, San Francisco, California, December 1, 1868

This ingenious device attempted to harness the energy of wave action for propulsion. A box or tank suspended below the deck was loaded with freight or ballast. As the sea pitched, it swung the ballasted box, actuating a propeller on the stern of the ship. Gearing, along with a specialized coil spring set between the power shaft and propeller shaft, equalized the action of the power on the propeller.

MARINE SAFES, PATENT NO. 180,575
Jean L. Gouley, New Orleans, Louisiana, August 1, 1876

In the unfortunate event that a sea vessel was in danger and the need for abandonment unavoidable, this new marine safe could store any gold, bonds, or jewelry, keeping them afloat should the ship be lost. The safe's hollow body, similar to a buoy, had hermetically sealed air compartments for the valuables. Its body was surrounded by cork, rubber, or other buoyant material, and the exterior was painted red for visibility.

LOCATING AND RAISING SUNKEN VESSELS, PATENT NO. 13,463
Joseph Hyde, New York, New York, August 21, 1855

This method for recovering sunken ships or cargo relied on detachable buoys that would float to the surface of the water while remaining attached to the vessel or cargo by a rope. Each buoy had coiled rope on a reel on its center, intended to unspool as it floated to the surface. Using a pulley attached to the buoy holder on the ship, a hook could be fed down using the surfaced rope to connect to and then raise the vessel or cargo. The buoy could also be used with an air hose connected to an inflatable bag, which could be filled with air to lift the vessel or cargo to the surface.

ROWLOCKS, PATENT NO. 168,000
John S. Dougherty, South Bay City, Michigan, September 21, 1875

Rowlocks hold the oar of a boat in place for rowing. This new rowlock facilitated the "feathering" of the oar. Feathering is a part of the rower's stroke where they turn the oar blade from being perpendicular to the water after the completion of a stroke to parallel with the water while bringing the blade back to begin another stroke. A specialized collar and sleeve design helped to guide the sweep of the oar allowing the "feathering" action automatically.

MODE OF CONSTRUCTING PADDLE WHEELS FOR PROPELLING STEAM AND OTHER BOATS, PATENT NO. 326
William A. Douglas, Albany, New York, July 31, 1837

This paddle wheel design kept the buckets or paddles of the wheel vertical while passing through the water, using levers, slides, and rollers to lock the paddle in position when entering the water and then allowing it to move freely when exiting the water. The wheel was 16 feet in diameter, and could accommodate 17 arms and paddles.

NOTE: This is the earliest numbered patent model in the "Rothschild Patent Collection."

SHEATHING FOR IRON VESSELS, PATENT NO. 54,083
Frederic Pelham Warren, East Court, Gosham, Great Britain, April 17, 1866

This innovation added an outer layer of copper sheathing to the iron plates of a vessel, sandwiching a layer of insulation between. By using a particular riveting technique and by separating the metals with insulating material attaching the copper plates to the iron sheathing, corrosive galvanic reaction between the two metals was minimized.

SELF LEVELING BERTHS FOR SHIPS, &C. PATENT NO. 193,837
John Calvin Thompson, Brooklyn, New York, August 7, 1877

This fanciful invention was meant to alleviate seasickness. Cots, seats, and lounges were connected with various levers, shafts, and rods that were ultimately connected to a large controlling weight located inside the ship. The inertia of that controlling weight would keep the berths level at all times, regardless of how rough the sea became.

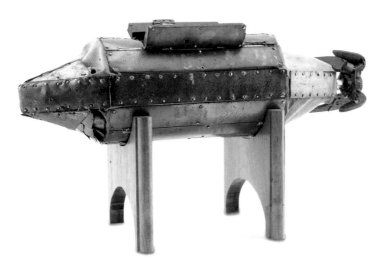

LIFE BOAT, PATENT NO. 31,562

John T. Scholl, Port Washington, Wisconsin,
February 26, 1861

This folding lifeboat's oblong slats of wood or metal were jointed and
hinged together to allow it to fold up. The passenger carriage within
it was mounted on rollers. The hull and carriage were independent
from one another and could complete revolutions in high seas.
Ventilation, steering, and a hand-powered propeller were provided.

MEANS FOR PROPELLING VESSELS, PATENT NO. 150,915

Hugo B.E. Von Elsner, St. Louis, Missouri, May 12, 1874

This invention improved the operation of reciprocating oars or paddles
for canal boats. Its slotted guides and adjustable collars for the paddle
levers kept the levers vertical while vibrating, adjusted the leverage of
the paddles, and also set the depth at which they operated.

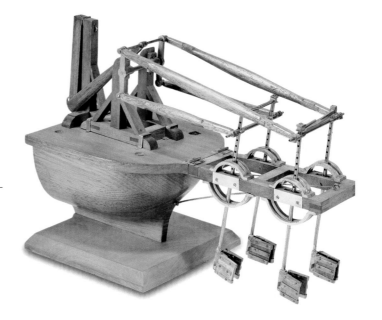

CHAPTER 16

Agriculture

"When tillage begins, other arts follow. Farmers, therefore, are the founders of human civilization."

— Daniel Webster

America was originally a nation of farmers. The Industrial Revolution of the 19th and 20th centuries transformed farming, as operations previously done by hand were increasingly turned over to machines.

Beginning in the 1840s, the invention and manufacture of a broad range of threshers, harvesters, binders, and other farm machinery, not to mention steam tractors, reduced human labor and increased productivity and yields.

As railroads revolutionized land transportation, they also changed American agriculture. By linking previously distant and isolated regions, the railroads enabled regions to specialize in particular agricultural products, as food could be shipped quickly over long distances. The growth of the corn and wheat belts from the Midwest to the Great Plains, the fruit and produce industries of California and Florida, and the beef industry of the Great Plains, all established their dominance at this time.

The rapid growth of the American population and the western expansion across the great frontier led to the establishment of many more American farms. Clearing the land was a major undertaking for farmers.

The 1862 Homestead Act hastened the settlement of the West. It granted adult heads of families 160 acres of unappropriated Federal land in exchange for a minimal filing fee and a commitment to five years of continuous residency on that land.

Thomas Jefferson, although responsible for numerous inventions, never took out a patent for his ideas. Jefferson believed in the natural right of all people to share useful improvements, without restraint. As Secretary of State, and the first supervisor of the Patent Office, he personally opposed patents. He firmly believed that invention should be for the good of all, not for the personal advancement of the inventor.

The patent system created under his tenure remains the basis for today's system.

While in Europe, as Secretary of State, Jefferson was exposed to the Dutch moldboard (the front of a plow that lifts up and turns the soil). He found the design clumsy and not as effective as he knew it could be. Jefferson would eventually invent a new one based on the basic principle of the right angle. His moldboards were widely used until the wooden plow was eventually upgraded to iron.

Between the 8th and 18th centuries, the tools of farming basically remained the same with few advancements being made. But increased market demand for agricultural products, from the 19th century onward, inspired many improvements in technology and a subsequent increase in farm production.

The first significant improvements in the machinery of farming were made to the plow, a basic farm implement that uses one or more blades to break up the soil and cut a small ditch (furrow) for the sowing of seeds. The cast iron plow, introduced around 1800, replaced the wooden moldboard plow. In 1837, John Deere's steel plow replaced the iron plows found across America's prairies. Deere was a blacksmith who moved from Vermont to Illinois in 1836. There, he discovered that the heavy, sticky prairie soil of the Midwest caked up iron plows, so he developed the first self-scouring plow, first using a steel sawmill blade. This new plow was able to shear cleanly through the region's rich, black soil. Deere sold his plows as fast as he could make them. As iron and steel became less expensive, and of better quality, moldboards were produced of stronger steel. Wheels were eventually added to plows. Lighter, better balanced, and more field-manageable plows began to be produced.

Mowing, threshing, haying machines, and cultivators,

were widely adopted in the two decades before the Civil War. Grain drills (which plant seeds in rows and then cover them with dirt) took the place of hand planting about the time of the Civil War.

Horse drawn mechanical reapers came into use in the 1830s, eventually replacing sickles for the harvesting of grains. Reapers developed into and were replaced by the reaper-binder (which cut the grain and bound it into sheaves), then came the combine harvester, in the mid-1800s, which simultaneously cut and threshed (separated the grain from the straw).

Fencing had typically been constructed using readily available natural materials, such as rock, wood, or thorny shrubbery. Such methods were problematic in the American West, as rocks for walls were scarce on the prairie, as was lumber for building rail or board fences. Thorn hedges took too long to grow and were, at best, only a solution for fencing small amounts of acreage.

The invention of barbed wire was a boon to farmers and ranchers, meeting their two most critical needs. Barbed wire was capable of restraining cattle, and it was also affordable. Consisting of wire strands twisted together with small pieces of sharply pointed wire spaced at short intervals, barbed wire allowed the effective fencing of large areas. Barbed wire was everything the typical fencing materials and methods of the day were not— cheap, quick, easy to install, and long-lasting. This was a fencing technology ideally suited to the wide open spaces of the western frontier.

Upon its introduction in the 1860s, the use of barbed wire quickly became widespread.

The adoption of it dramatically changed life on the frontier. Land and water, once available to all, was now claimed and fenced off by ranchers and settlers. Barbed wire also found an application during war, with dense, twisted coils of it used to thwart the advance of enemy armies and to protect people and their property. Today, barbed wire fencing remains the standard for enclosing cattle in most regions of the U.S.

CYRUS MCCORMICK (1809-1884)

Cyrus McCormick was the son of Robert McCormick (who was also the inventor of a threshing machine). In 1831, Cyrus built the first practical grain harvesting machine, or reaper. His machine contained the essential components found in every grain harvester of the time, and even to the present day.

McCormick's reaper consisted of short triangular knives with small cutting edges that were serrated with sharp ridges. McCormick's reaper radically increased production. It could cut 15 acres of wheat in a day while a man with a scythe and cradle could cut only about three acres. McCormick's machine proved to be unrivaled. As a result, his reaper was awarded a medal at the London Crystal Palace Exhibit in 1851.

Cyrus was an inventor, manufacturer, financier, sales and public relations man. He moved west and founded his factory in Chicago to be nearer to the demand for his invention. In the 1850s, he was one of the first businessmen to locate his agents throughout the land, develop exclusive dealerships, and sell machines on credit.

DANIEL MARMON (1844-1909)

Daniel Marmon was orphaned at age five and raised by his uncle. After graduation from college, Marmon went to work at Nordyke, Ham & Company, a mill construction business in Richmond, IN, founded in 1851 by Ellis Nordyke. In 1866, Marmon became an equal partner in the company and it was rechristened Nordyke, Marmon & Company.

By 1871, the company was one of the most prominent businesses in its industry and was known as "America's Leading Mill Builders." Eventually, Nordyke, Marmon & Company would export machinery throughout the world.

The company offered complete machinery for flour, corn, cereal, and rice mills, and for elevators. They produced such equipment as roller mills, bolting machines, packers, blending machines, and rice, corn, and starch milling machines. (Figure 1)

Daniel Marmon also owned Noblesville Milling Company of Noblesville, IN, and was president of the Indianapolis Light and Heat Company.

Figure 1: Mill Sheller, Nordyke & Marmon Co.

In 2008, Berkshire Hathaway Inc. acquired majority interest in The Marmon Group of Companies. Still in business today, it is a conglomerate of manufacturing and service businesses.

SHELLS FOR INCASING MILLSTONES, PATENT NO. 216,048
Daniel W. Marmon, Assignor to Nordyke & Marmon Company, Indianapolis, Indiana, June 3, 1879

The cylindrical shell of this portable mill is constructed in two parts or rings. The lower ring forms a shallow shell, in which the running stone is mounted. The other ring fits into the lower shell, extending the shell above the upper face of the running stone. This new construction makes it easier to indicate irregularities and to obtain a level and true face of the stone.

MEINRAD RUMELEY (1823-1904)

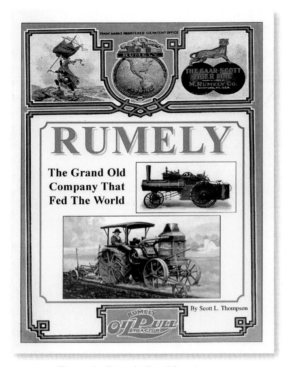

Figure 2: Rumely Co. Advertisement

Meinrad Rumely was born in 1823, in Adelsburg, Germany. He arrived in New York in 1848 and worked in the midwest in a variety of mechanical-related jobs before settling in La Porte, IN where he opened a blacksmith shop with his brother, John. The brothers set their sights on creating and improving farm machinery.

In 1854, M. & J. Rumely entered the farming equipment business, introducing their first thresher. The machine won first prize at the Illinois State Fair in Chicago five years later. The company went on to capture a large portion of the thresher business in the U.S. and abroad.

In 1887, Meinrad became president and general manager of the company, then renamed the M. Rumely Company. The company produced an improved plow, separators, stationary engines, boilers, traction engines, spark arresters, friction clutches, ice elevators, and other machinery and appliances. By 1882, M. Rumely Company was building steam traction engines (a.k.a. tractors). In 1886, the company introduced a new straw-burning steam engine. Rumely's prize-winning thresher was one of the earliest to be powered by steam. During its heyday, the M. Rumely Company was the largest manufacturer in the city of LaPorte, IN. (Figure 2)

SPARK ARRESTER, PATENT NO. 218,643
Meinrad Rumely, La Porte, Indiana, August 19, 1879

The "Rumely Spark Arrester" prevented sparks and other flammable debris from escaping from the internal combustion engines of farm tractors. Its design included a cylindrical screen at the top of the stack, along with the conical section, to impede the draft and contain the sparks.

SYLVANUS DYER LOCKE (1833-1896)

Figure 3: The Locke Wire Binder of 1873

Sylvanus Locke was born in Richfield, NY, the youngest of 11 children. His self-binding harvester (Figure 3) included a binding device that bound, cut and raked grain, and produced it into bundles. It was the first automatic self-binding harvester to be sold commercially. Locke's binder ended up being adopted by Cyrus McCormick.

Locke received nearly 50 patents relating to harvesters and binders. He served one term in the New York state legislature (1883-1885) and also invented a train car coupling device and a machine for manufacturing detachable steel link belting.

HARVESTER, PATENT NO. 214,929
Sylvanus D. Locke, Egbert Bowhay,
Hoosick Falls, New York, April 29, 1879

This grain conveyer for harvesting machines used spiraling chutes to avoid clogging. Their discharge ends were "entirely free and unconnected with any internal rotating shaft or fixed support." This allows for a smooth and unobstructed delivery of the grain and the conveyer is prevented from becoming clogged and rendering the machine inoperable. A second series of spirals on the grain platform may also be used for elevating the grain.

CORN STALKER CUTTER, PATENT NO. 229,106
Ezra Dominy, Freedom, Illinois, June 22, 1880

This corn stalk cutter includes a replaceable bearing for the cutter arms, minimizing wear on the crank axle and the outer surface of the tubular bearing, resulting in little resistance by the moving parts, and simplifying lubrication. When moving the machine between fields, the cranked portion of the axle can be raised to protect the blades.

MACHINE FOR CUTTING GREEN CORN OFF THE COB, PATENT NO. 256,926
Welcome Sprague, Farnham, New York, April 25, 1882

This machine combines a feed belt, cutters, scrapers, feed wheels, and connecting parts to maximize its cutting and scraping capacity. The kernels of the corn are completely removed before the cob is discharged from the machine. The manner in which the ears are driven against the cutters has also been improved resulting in less damage to the cutting mechanism.

SHEEP FEEDING RACK, PATENT NO. 353,384
Denzil Phillips, Cochranton, Pennsylvania, February 2, 1886

A combination of parts and construction provide for a device that may be readily and quickly taken apart and folded, and which will occupy a minimal amount of space. It also allows for a feeding rack whereby it also prevents the animals from wasting their feed. The inclined sides of the rack feed the hay to the bottom of the rack.

BARN DOOR FASTENING, PATENT NO. 365,649
Clarence C. Sprinkle, Majenica, Indiana, June 28, 1887

This fastening device allows a barn door to be kept open or shut. To keep the door open, the bar is secured into the door jamb. To keep the door shut, the bar is secured to the hook on the opposite door jamb of the hinged door. This fastening device can be used with any type of barn or similar door.

BALING PRESS, PATENT NO. 214,167
Edmond F. McGowen, Houston, Texas, April 8, 1879

This invention speeds the baling of cotton or other materials. The material is placed in a press box, and then compressed by a screw follower operated by horse or other power. Reversing the screw withdraws the follower without stopping or changing the motion of the driving shaft. This improved mechanism reduces the baling time compared to other baling machines of this period.

CULTIVATOR, PATENT NO. 31,367
Charles Beach, Thomas Brown, Jacksontown, Ohio, February 12, 1861

Wrought iron is used to construct a plow for the cultivation of Indian Corn and other plants. The frame is in usual form, except that the end of the beam is curved and pierced with two or more holes for changing the draft. Adjustable holes allow for the handles to be laterally adjusted. The shovels are slightly concave and attached to the beam by bolts. The shovel is constructed to throw the earth up to the plants when the plants require hilling.

CATTLE CHUTE, PATENT NO. 221,931

James T. McCoy, Robert A. McCoy, McCoy's Station, Indiana, November 25, 1879

This adjustable chute simplifies the loading of cattle into a railroad car. When not in use, it can be raised out of the way of the moving train cars. The chute is raised by the use of straps, ropes and pulleys. The weights balance the forward end of the chute, and the slatted floor allows dirt to fall through and not hinder the operation of raising and lowering the platform.

APPARATUS FOR SEPARATING BURGEONS FROM GRAIN, PATENT NO. 186,328

Charles A. Duprez, Reims, France, January 16, 1877

This apparatus removes "burgeons or sprouts, germs or other foreign bodies" from stored grain. The back and forth, and up and down motion of the brushes rids the grain of sprouts and other impurities. The cleaned grain is then removed by passing down the inclined sieve.

SHEEP SHEARS, PATENT NO. 52,293

Albert H. Kennedy, Brunswick, Ohio, January 30, 1866

This tool for shearing wool from sheep was called the "Kennedy Sheep Shearer." The application of power was transmitted through spiral spring wires so that sheep could be shorn with great facility and ease. A band running over the driving wheel of the machine had a spiral spring attached to the arm. Attached to the spring were the shears, which have a flexible horizontal motion to cut the wool.

GRUBBING MACHINE, PATENT NO. 218,127
Shubael G. McCann, Corvalis, Oregon, August 5, 1879

Grubbing is the process of removing vegetation from the ground. This machine consisted of a wooden frame with two solid iron rollers cast upon a shaft with bevel gear pinions. The master wheel rested upon a cast iron seat, and the wheels had bent axles in the shape of a crank, so the machine could be lowered to the ground. This new and improved grubbing machine is easily adapted for the work to be done with little loss of time and labor.

HARVESTER, PATENT NO. 81,241
George W.N. Yost, Corry, Pennsylvania, Assignor To Corry Machine Company, August 18, 1868

The uniqueness of this harvester is its new and improved "Climax Body for Incasing and Protecting the Gearing of Grass and Grain Cutting Machines." Its casing could contain and protect the gearing of any harvester.

BARBED FENCE WIRE, PATENT NO. 199,538
John Hallner, Ithaca, Nebraska, January 22,1878

A narrow strip of sheet metal forms a series of barbs. The strip is then twisted around an ordinary fence wire, allowing the barbs to stick out in all directions. The barb wire, inexpensive to produce, is light and durable, and will prevent stock from breaking through fences erected with it.

PACKING BOXES FOR EGGS, PATENT NO. 215,933
Ignatz Karel, Blue Earth City, Minnesota, May 27, 1879

Eggs need to be protected when being packed and transported over long distances. This egg box design used alternating long corrugated springs and plain strips of wood, on the box bottom, and topped that with a wooden plate containing cylinders to hold the eggs. Layering more plates and cylinders would accommodate more eggs. Finally, the last cylinder was topped with a plate with a lockable box cover.

MACHINE FOR MAKING HARROW TEETH, PATENT NO. 240,920
James Morgan, Pittsburg, Pennsylvania, May 3, 1881

Harrow teeth are the sharp tines that break up clods of soil, remove weeds, and cover seeds as the harrow is dragged over plowed land. The machine was designed expressly for making harrow teeth having a head and a four sided point. It employed feed rolls which rotated continuously, giving a forward feed to the bars. As they fed forward, they severed, or partially severed, the successive blanks, giving them a chisel or wedge point at one end.

GRAIN REGISTER, PATENT NO. 186,042
Milton W. Nesmith, Metamora, Illinois, January 9, 1877

This register operates using a rotary table, which is turned by passing the measures of grain on it. The half-bushel measures of grain are placed upon the table, then removed as soon as it has been registered. The register could record one thousand bushels and then reset to begin again.

CHAPTER 17

Tools

"A tool is but an extension of a man's hand, and a machine is but a complex tool and he that invents a machine augments the power of a man and the well being of mankind."

— Henry Ward Beecher

There was no tool more important in shaping the American frontier than the axe. Axes built shelters, cleared land, and prepared fuel. The axe was one of the first basic tools to take on a new form in the U.S. The American axe was unique because the bit of the axe (the cutting edge), was about the same weight as the poll (the flat edge, butt), giving it excellent balance. The European axe had a longer, narrower bit, and hardly any poll at all. This difference allowed the American axe to be swung straight and clean. An American axeman could fell three times as many trees in the same amount of time as someone wielding a European axe.

Other early tools considered essential in turning the wheels of progress included the lathe, the planer, and the milling machine, all considered essential machine shop tools in mid-19th century. From 1800 on, the lathe was essential for iron and steel work.

A craftsman's tools were his main means of earning a living. His tools were either made by the craftsman himself or purchased from specialty manufacturers and suppliers. The cost of a fully-equipped tool chest in the late 18th century was about a year's wages for a skilled craftsman. Throughout the 18th century, wooden tool components were gradually replaced with metal ones. By the middle of the 18th century, an all-metal lathe had been constructed.

The invention and development of machine tools was essential to the success of the industrial revolution. The steam engine, the railroad, and the textile industry required machine tools for their operation. In the second half of the 19th century, machine tool construction benefited greatly by the increasing use of interchangeable parts, improvements in cutting tool materials, the increased demand for small arms as a result of wars in Europe and America, the introduction of new inventions, and the industrial application of electrical power. The commercial or technical success of new inventions were often delayed until improved machine tools were developed to actually realize them.

Eli Whitney, the inventor of the cotton gin, established a gun factory in Connecticut, and in 1818, invented a milling machine for producing rifle parts. The Civil War brought about a need for greater production output with less labor requirements. This contributed to the evolution of automatic machine tools. Automatic lathes for the mass production of screws were built during the war. As discussed previously in Chapter 3, Alcohol, Tobacco and Firearms (ATF), early gun makers desired interchangeable parts for their guns that could be swapped in the field. The use of machines reduced the labor-intensive cost of producing guns in quantity. During the Civil War, a great numbers of guns were needed on both sides and screws were an essential part of producing these weapons. The turret lathe was developed during this period, which eliminated the need to change tool settings for each screw.

The U.S. took the lead in developing woodworking machinery, in part because trees were so plentiful from sea to shining sea. The move westward also created demand for more building materials and home furnishings. As early as 1820, doors, window sashes, room furnishings, spinning wheels, gunstocks, and many types of clocks were being machine-manufactured. As a result of this acceleration of activity, most of the specialized processes for working wood by machine originated in the U.S.

At the 1876 Centennial Exhibition, some of the most impressive of all of the exhibits were mounted by the manufacturers who showed a supply of tools designed to saw, shape, mold, mortise, tenon, or turn wood, all driven through overhead shafts and belts turned by George Corliss' massive steam engine. Pratt & Whitney Company of Hartford showed 49 different varieties of machine tools. Multiple spindle drills, milling machines, bolt cutters, boring mills, power shears, fire arms, and metallic cartridges were all at the Centennial Exhibition and American firms dominated the awards.

As the U.S. moved from an agricultural-based society to an industrial one, it was the development of machine tools that made this change possible. The textile industry with mechanical looms, steam engines, the railroad industry, clocks and watches, sewing machines, and guns, all required metal parts in greater quantities made more efficiently and with greater accuracy. The origins of the American machine tool industry shows the importance of the machine tool in the development of American manufacturing. Mass production, standardization of parts, and precision manufacturing are all owed to the use of machine tools.

LAROY SUNDERLAND STARRETT (1836- 1922)

Laroy Starrett invented the combination square in 1877. In 1880, he founded L.S. Starrett Company, Athol, MA to manufacture his invention. L.S. Starrett is still located in Athol and is still the company's world headquarters. In the 19th century,

Figure 1: Starrett Factory

the plant was the largest in the world devoted exclusively to the manufacture of small tools and hacksaw blades. Starrett was president of the company from 1880 to 1922.

Laroy S. Starrett was born in China, ME, one of twelve children to farmer parents, Daniel D. and Anna Elizabeth Starrett. While working on his farm, Laroy invented a washing machine and a new type of butter churn.

Starrett both invented new tools and techniques as well as made improvements to existing products. His first invention was a machine to chop meat which he called the "Hasher." Not only did Starrett manufacture the Hasher, but also

sold it and the patent rights to it throughout the state of Maine. Starrett entered into a contract with the Athol Machine Company to manufacture the meat chopper and some of his other inventions.

Starrett received over 80 patents for his inventions including a "center try square" designed to find the exact center of a circle. There were also many patents assigned to his company. He manufactured the combination square consisting of a try square with a moveable head that could be clamped in any desired position along the blade. Other tools included a surface gauge, a beveling instrument, a screw thread gauge, calipers and dividers, hammer, screwdriver, car coupling, wrenches, vise, ratchet wrench, screw holding screw driver, hacksaw, combined level, square and plumb, pliers, combined gauge and level, measuring gauge, expansion pliers and adjustable square. (Figure 1)

WRENCH, PATENT NO. 190,636
Laroy S. Starrett, Athol, Massachusetts, May 8, 1877

The adjustment to open and close the wrench was designed for the wrench to easily slip on a slide bar to the required spacing needed by depressing a thumb piece which disengaged the bottom jaw. The bottom jaw had a "half thread or half nut" that mated with an internal screw which was found inside the body of the slip bar. Fine adjustment to the jaws of the wrench were made by turning a collar located at the bottom of the slip bar.

LINUS YALE, JR. (1821-1868)

Linus Yale Sr. invented a pin tumbler lock in 1848. His son, Linus Yale Jr. (Figure 1) improved upon his father's design using a smaller, flat key with serrated edges that is the basis for the modern pin tumbler lock. Linus Yale Jr. patented his lock in 1861. He invented the modern combination lock in 1862. His father was a locksmith and inventor who held eight patents for locks and another half-dozen patents for threshing machines, a sawmill head block, and millstone dresser.

Throughout his career, Yale Jr. received many patents, mostly related to his inventions of locks and safes, but also relating to mechanical problems. In 1858, he patented a device for adjusting the right angle of a joiners' square. In 1865, he patented a tool for reversing the motion of screw taps. He also received two patents for

Figure 2:Portrait of Linus Yale Jr.

improvements in mechanics' vises, in 1868. Yale Jr. also obtained foreign patents for some of his inventions.

As part of Yale's business plan, and to promote his bank locks, Yale challenged anyone to pick his locks and was so confident in their security that he offered a $3000 reward (a considerable sum at the time) if someone could successfully pick one of his locks.

Yale Jr. established the Yale Lock Manufacturing Company in 1868 in Stamford, CT with his business partner, Henry Robinson Towne. The Yale Lock Company became America's number one manufacturer of locks. Linus Yale Jr. forever changed the lock industry with his many breakthrough inventions, his most important one being the Yale Cylinder Lock.

ADJUSTABLE SQUARE, PATENT NO. 21,861
Linus Yale, Jr., Philadelphia, Pennsylvania, October 19, 1858

This new method for constructing squares has a novel way to readjust the squareness of the tool when it is purposely or accidentally misaligned. The lateral arm to be adjusted is attached to the straight edge by a rivet pin. The lateral arm has a recess in the wood frame. The recess is occupied by a blade that has some room to move within the chamber and at the end of this blade there are two fine adjustment screws that act on the blade and are turned until the lateral arm is brought back to square.

SAMUEL JOHNSTON SHIMER (1837-1901)

Sons of Abraham B. and Margaretta Johnston Shimer, Samuel Shimer spent his early years working as a farmer while his brother, George was in the lumber business. In October 1871, Samuel went to Milton, PA to work with his brother and the next year they established a saw and planing mill, Applegate, Shimer & Company, with George Applegate and C.L. Johnston. Their main business was manufacturing of lumber, but they also operated a small machine shop which was where, in 1873, George and Samuel invented a cutter head. In May of 1880, their business was destroyed in a fire. The brothers rebuilt the plant as a machine shop that was devoted to the manufacturing of cutter heads. Their cutter heads were sold throughout the U.S., Australia, England, and Canada.

In 1884, George Shimer retired from the business and Samuel became sole owner. He took his sons, Elmer and

Figure 3: Shimer Co. Advertisement

George, into the business, which was then renamed S.J. Shimer and Sons. In the fall of 1888, the Shimers assumed control of the Milton Manufacturing Co. In the spring of 1889, Samuel J. Shimer invented and patented a machine for cutting washers. These washer cutting machines were used in the Milton Manufacturing plant. While known primarily for their cutter heads, they also made planers and spindle shapers. In 1886, the company claimed their cutter heads to be the strongest, cheapest, and most durable cutter heads made. In the woodworking trade, the Shimer company's cutter heads for wood molding machines were the industry standard.

Between 1873-1920, at least 40 patents were issued to individuals working at Samuel J. Shimer & Sons (patents could not be issued in a company name).Of these, at least eleven were issued to George J. Shimer, and twenty four to Samuel J. Shimer. (Figure 3)

CUTTER HEAD, PATENT NO. 236,636
George J. Shimer, Samuel J. Shimer, Milton, Pennsylvania, January 11, 1881

This cutter head was used for matching and molding lumber and was particularly used where bits were employed to produce a divided cut and to operate with clearance at the side of each cutting. The cutter head had "T" shaped slots alternately inclined at the axial line of the head. The head was designed to hold segmental cutters arranged in the "T" slots and secured. The individual cutters were capable of independent adjustments and were to be repositioned accordingly to adapt to the different thicknesses of material to be worked.

TILE FACING AND SQUARING MACHINE, PATENT NO. 228,929

Charles F. Powers, Sutherland Falls, Vermont, Assignor to Sutherland Falls Marble Company, June 15, 1880

This machine could smooth the faces and edges of several tiles at a time. The tiles were held face down on a spinning rubbing bed or grinding plate, while the sides or edges were honed to the desired squareness. The machine operator could remove and replace a tile without affecting the other tiles.

MACHINE FOR EDGING, SIZING, AND STRAIGHTENING NON-CYLINDRICAL METAL BARS, PATENT NO. 209,588

Joseph S. Seaman, Pittsburg, Pennsylvania, November, 5, 1878

This machine could straighten large non-cylindrical metal bars such as railroad rails, girders, and beams. It held the bar firmly as a pair of rolls mounted on rocking frames forced the bar into a straight beam.

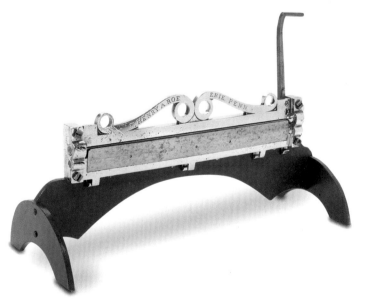

MACHINERY FOR BENDING SHEET METAL, PATENT NO. 4,187

Henry A. Roe, Erie, Pennsylvania, September 11, 1845

Until this invention, creating a complete fold in sheet metal required a separate, secondary process. This invention improved the folding process so two or more pieces of sheet metal could easily be linked together to form cylinders, roofing, and various other objects.

SET WORK FOR SAW MILLS, PATENT NO. 233,409

William Gowan, Wausau, Wisconsin, October 18, 1880

This set work allowed a sawmill operator to better direct the log into the saw blade. The set work's "dogs" could be individually adjusted against the log, and then all moved simultaneously using a hand wheel. The operator could see the adjustments by referring to the dials, which were designed to avoid the accumulation of saw dust.

SAW FILING MACHINE, PATENT NO. 213,925

Thomas L. Nanney, Evansville, Indiana, April 1, 1879

This invention improved the time-consuming process of saw sharpening. The machine held the saw in a vice while the operator slid the sharpening file forward and backwards. The file could be set to a desired degree or angle. To move to the next tooth, the operator moved the blade forward using a pivoted hand lever set to travel the required distance.

SOLDERING IRON HEATER, PATENT NO. 213,493

Josiah Burgess, Zanesville, Ohio, March 25, 1879

This invention placed a reservoir of heating oil and a burner plate below the chamber where the soldering iron rested. The burner plate deflected heat evenly, so the iron heated on all sides. The window for monitoring the flame doubled as a lantern.

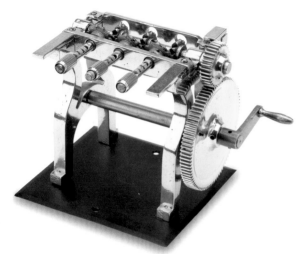

MACHINE FOR CUTTING CORKS, PATENT NO. 8,422
George Hammer, Philadelphia, Pennsylvania, October 14, 1851

Cork's peculiar softness and elasticity made it difficult to cut without excessive heating that destroyed the edges of the cutting tools used at the time. This machine's revolving cutting burr surrounded a cutting cylinder. The cutting cylinder first cleanly pierced the cork, and then allowed the cutting burr to follow, removing the surrounding cork to shape the cylinder into a bottle cork.

CUTTER HEAD FOR FRIZZING MACHINES, PATENT NO. 203,552
Rudolph Naatz, Berlin, Prussia, Germany, May 14, 1878

This invention improved the adjustment of frizzing (rotary) cutters for wood and metal working machines. The individual cutters on the hub could slide up and down, and be pivoted to the angle needed as the cutter moved forward or in reverse. All the cutters could be moved as one, making adjustments easier.

SHEARING MACHINE, PATENT NO. 84,695
Samuel W. Huntington, Augusta, Maine, December 8, 1868

These shears were designed for tailors use for cutting thick fabrics, as well as metal workers cutting thin metals and wire. The user could adjust the power by adjusting the position of the fulcrum. At the rear of the machine, auxiliary blades could cut thick wire and other hard materials.

PARALLEL RULER, PATENT NO. 14,396
R. Eickemeyer, Yonkers, New York, March 11, 1856

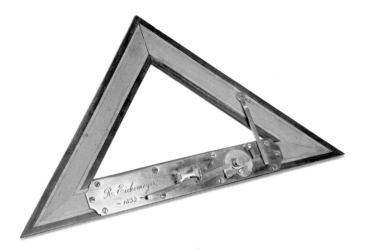

This ruler allowed a draftsman to graduate the distance of ruled lines by pressing a small lever to engage a "foot piece" that allowed the ruler to shift by the desired distance. The ruler could also guide the peripheries of specialized cams or patterns that can be utilized to rule irregular or progressively graduated distances, especially when representations of cylinders or other bodies are ruled.

SCREW DRIVER, PATENT NO. 147,059
George P. Loomis, Utica, New York, February 3, 1874

Instead of turning the screwdriver by hand, the user of this screwdriver pushed on the screw, and the screwdriver blade turned, using an internal double toothed pawl and disc catch. When the user stopped pushing, the pawl slipped over a projection on the disc, disengaging the turning action of the blade. To reverse the action of the screw driver, the operator switched the pawl to the other end of the slot, bringing the opposite side of the pawl in contact with the disc catch.

PIPE CUTTER, PATENT NO. 89,959
Ambrose G. Wilder, Cohoes, New York, May 11, 1869

This invention cut pipe with minimal effort. It combined a solid cast iron frame with a knife cutter and rollers to support the pipe as it was cut. A coil spring between the handle and the knife allowed a small amount of "give" if the pipe was irregular in shape. Turning the handle applied pressure to the pipe, and the pressure gradually increased as the cutter spun around the pipe until it was severed.

TOOL FOR CUTTING OFF BOLTS, RODS, & C., PATENT NO. 104,061

Henry Peters, Davenport, Iowa, June 7, 1870

This new bolt cutter reduced the amount of effort needed to cut bolts or rods. Its blades were beveled inwards to produce a flush cut of the rod or bolt against the nut or surface. After the cut was made, a spring set between the levers assisted in opening the levers.

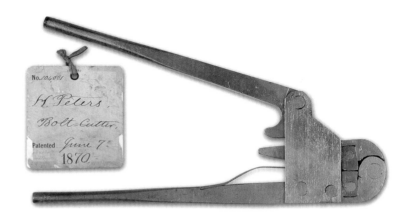

DRILL BIT, PATENT NO. 124,089

Henry S. Shepardson, Shelburne, Massachusetts, February 27, 1872

This drill bit shank or holder could hold interchangeable bits or cutters of various sizes and forms. Its novel means of attaching the bit to the shank secured the bit firmly in place and prevented any side motion while in use. The bits' spurs were unusually wide and long, giving them great strength at a common weak point and a longer working life.

HOLLOW AUGER, PATENT NO. 64,478

George E. Booth, Seymour, Connecticut, May 7, 1867

This tool was designed primarily to produce the tenons for wagon wheel spokes. Its auger revolved in a lathe or hand brace, and could accept a "thimble" to vary the diameter of the tenon being cut. The cutters on the auger are made to adjust to accommodate the various thimble inserts.

BENCH VISE, PATENT NO. 27,960
Levi A. Beardsley, South Edmeston, New York, April 24, 1860

This vise was strong and very adjustable. Its jaws were strongly pivoted to two arms at their lower ends, and were tapped to accept a turning screw so they could open and close with a handle. In the middle of the turning screw, a small wheel turned in a grooved projection of a guide bar to keep the bar in place. The arms secured the vise to a bench or table. By reversing the position of the screw and guide bar, the vise could be used as a wrench.

RATCHET DRILL, PATENT NO. 116,847
James W. Mahlon, Brooklyn, New York, July 11, 1871

This ratchet is an improvement on the tools that require a ratchet action, such as wrenches, drills, etc. The improvement lies in the construction of the ratchet wheel and pawl in combination with the handle. The novel polygonal ratchet wheel design had rectangular recesses and a lever with projecting rectangular pawl teeth, one on each side for reversal settings.

CHAPTER 18

Miscellaneous

*"One can be absolutely truthful and sincere even though admittedly the most outrageous liar.
Fiction and invention are of the very fabric of life."*

— Henry Miller

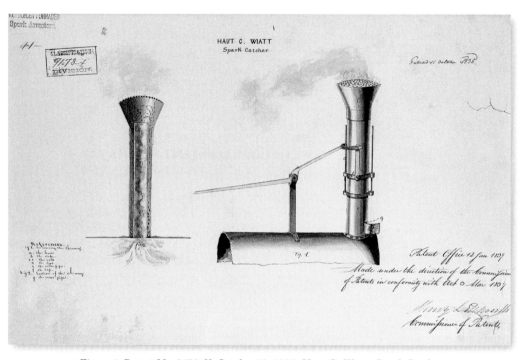

Figure 1: Patent No. 9173-X, October 15, 1835, Haut C. Wiatt, Spark Catcher

PATENT DRAWINGS

Patent drawings had been required from all applicants applying for a U.S. Patent since the first patent statue was enacted in 1790. The fire of December 15, 1836 destroyed the entire Patent Office located in the Blodgett Hotel in Washington D.C. along with all of the records, drawings, and most of the patent models. Between 1790 and 1836, the Patent Office granted 9,902 patents that were referenced only by the inventor's name and date. After the fire, approximately 3,000 models and drawings were recreated and were marked with an "X" suffix to distinguish them from the new patents issued after July 4, 1836, that were numbered starting with the number 1. (See Figures 1, 2, and 3). The drawings from 1837- 1870 are, for the most part, the original drawings submitted by the inventors. (Figures 4 and 5)

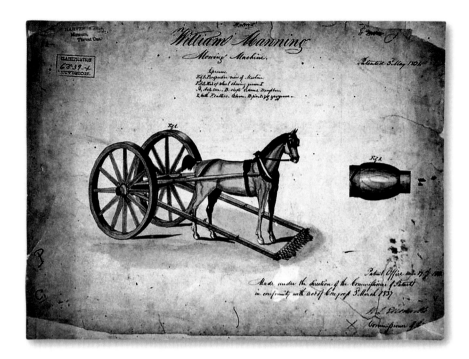

Figure 2: Patent No. 6593–X, May 2, 1831, William Manning, Mowing Machine

Prior to 1870 the Patent Office had not set forth any specific requirements as to the size or the medium to be used for the artwork of the drawings. Many drawings from this period were very elaborate, done by professional artists using oil paints and water colors.

Beginning in 1871, the Patent Office required that all drawings submitted for a patent application were to be black on white and of a specific size. The Patent Office hired the Graphic Company of New York, NY to redraw all the patent papers from 1790-1870 to conform with the new requirement for drawings initiated in 1871. (Figure 6)

Figure 3: Patent No. 6490–X, April 18, 1831, James Johnson, Fire Ladder

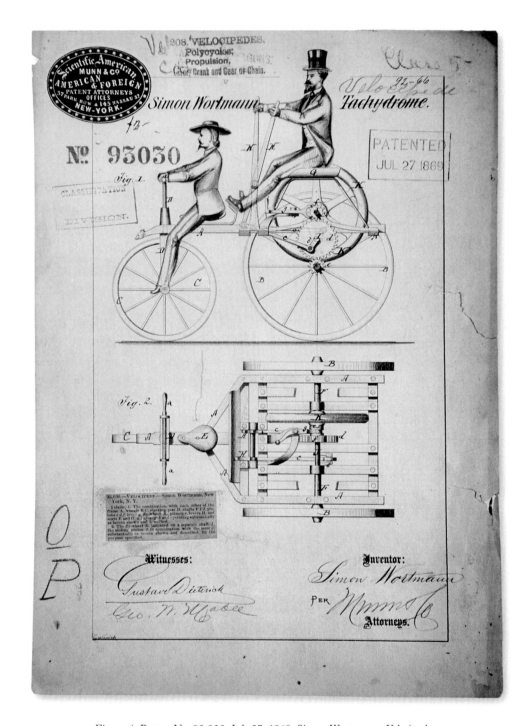

Figure 4: Patent No. 93,030, July 27, 1869, Simon Wortmann, Velocipede

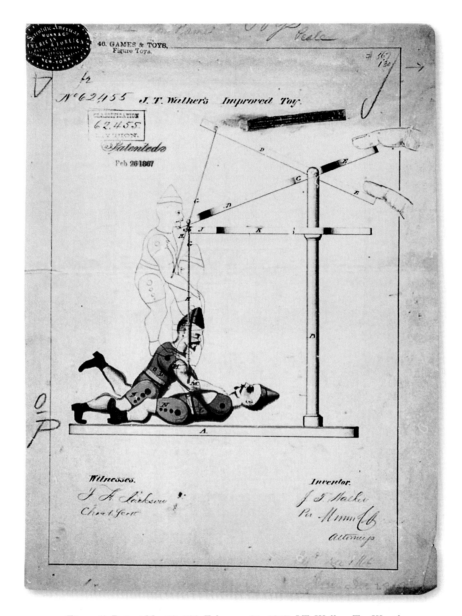

Figure 5: Patent No. 62,455, February 26, 1867, J.T. Walker, Toy Wrestlers

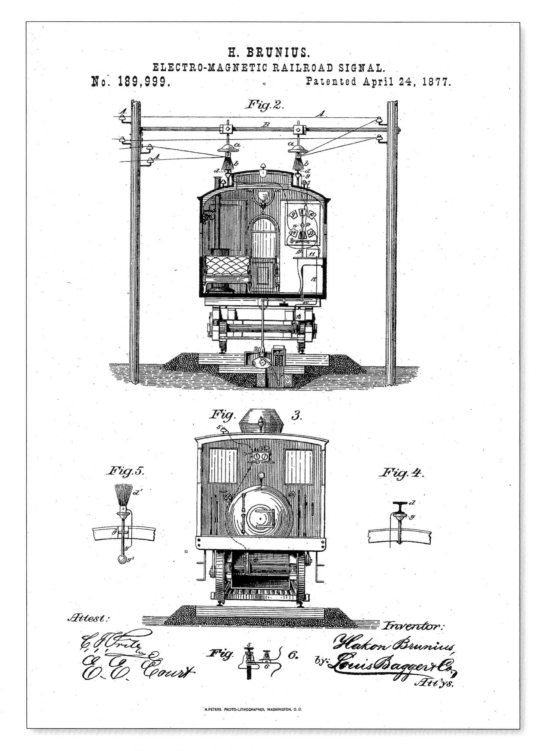

Figure 6: Example of uniform black on white patent drawing

PATENT TAGS

During the patent model era of the United States Patent and Trademark Office, the Patent Office affixed a patent and a receiving tag to each patent model granted a patent. The patent tag (3" x 3") was attached to the model by a short length of red twill tape. The expression "Government Red Tape" may have its origin in this practice. The receiving tag (1½" x 3") was attached to the model when the inventor submitted the application and model to the Patent Office. (see Figure 7)

From 1790-1836, the patent tag had only the inventor's name, the invention name and date, and no patent number on the tag. (See Figure 8). From 1837 until patent models were no longer required, the patent tags had the addition of the patent number in the top left hand corner.

In 1842, the U.S. Patent laws were modified to allow patenting for a variety of different designs. A separate numbering system was started with Design Patent Number 1 being issued to George Bruce on November 9, 1842 for a new typeface. The new patent tag was similar but had the addition of "Design for" written on the tag. (See Figure 9)

In 1876, to celebrate the 100th anniversary of the signing of the Declaration of Independence, the U.S. Patent Office issued a tag in a very limited amount to applications that received a patent, with the year 1776 printed in blue and year 1876 printed in red on the white background of the tag. Also, a red, white, and blue ribbon replaced the red tape in just a handful of the models that had the special red, white and blue 1876 tag. (See Figure 10)

On September 24, 1877, a second fire occurred in the Patent Office in Washington that had supposedly been built to be fire-proof. Approximately 80,000 models were destroyed and possibly as many as 60,000 models were rescued from the debris of the fire. These models were signified by the issuance of a new tag that stated on the back of the tag that it was taken from the fire. (See Figure 11)

Over the years many patent tags have become separated or lost from some of the models. Without this patent tag, confusion sometimes arose as it has become difficult to identify whether these were original patent models or salesmen samples, which were also prevalent during this period of time.

Figure 7:
Typical patent tag and receiving tag
from 1837 until end of model requirement

Figure 8:
Example of Patent tag,
1790-1836, with no patent number

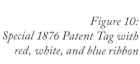

Figure 9:
Example of Design Patent Tag, from 1842 until end of model requirement

Figure 10:
Special 1876 Patent Tag with red, white, and blue ribbon

Model taken from the debris of the fire of Sept. 24, 1877.

Figure 11:
Example of patent model tag that survived the fire of 1877

TELEGRAPHIC APPARATUS, PATENT NO. 29,761
L. Bradley, New York, New York, August 28, 1860

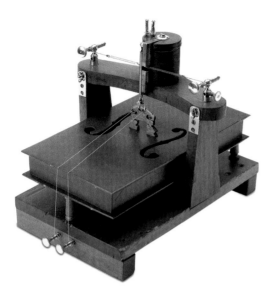

By increasing the the power of the electromagnets, it gives greater freedom and ease to the movement of the armature and increases its capacity to make a greater number of contacts in a given time. The helix is arranged so it can increase or diminish the power of the magnet as required and to give greater distinctness to the sound produced by the strikes of the armature in its vibrations. The helix is composed of series of concentric sections of insulated wire, through the center of which passes a core, or magnet, of soft iron.

VENTILATING BUILDINGS, VESSELS, &c., PATENT NO. 216,804
Francis L. Norton, New York, New York, June 24, 1879

The apparatus consists of a system of pipes connected with apartments to be cleared of foul air, and communicating through a clapper valve and a sliding pipe. An exhausting and forcing device, consisting of a water bell, to which a vertical reciprocating motion is imparted, is operated by a steam or water engine or other motor. The return or down stroke of the bell forces the air through the sliding pipe, and is forced through the clapper valve opening in the opposite direction, and discharged into the open air.

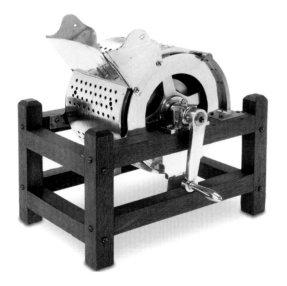

OATMEAL MACHINE, PATENT NO. 207,152
George Ayliffe, Akron, Ohio, August 20, 1878

Hulled kernels of oats are cut by cutting or shearing edges. The machine rapidly cuts the kernels of oats into uniform particles of any desired size. A revolving hollow cylinder is perforated with a series of holes, into which, upon the the upper side of the cylinder, the grains fall endwise, and while standing are carried against the edge of the knife and severed.

MOUSE TRAP, PATENT NO. 102,133
John O. Kopas, George W. Bauer, Washington, District of Columbia,
April 19, 1870

This is a humane trap that can capture up to four live mice. There are four trap doors that are provided with a bait holder that are all connected with a latch mechanism to a common shaft. The trap door opens when the mouse takes the bait, the mouse falling into the box, followed by the next baited platform rotating into place. The mice are removed from the bottom of the trap by a sliding door.

Ralph Waldo Emerson, the great essayist, said in a lecture in 1871, "If a man can write a better book or preach a better sermon or make a better mousetrap than his neighbor, even if he builds his house in the woods, the world will make a beaten path to his door." 4,400 inventors tried making a better mousetrap and received patents from the U.S. Patent Office. Only about 25 were ever profitable. In 1895 John Mast of Lancaster, Pennsylvania, invented the "snap trap." It was very effective, simple in design and cheap to manufacture. Today the Woodstream Co. manufactures this trap under the Victor name and sells over 10 million a year, which is over 60% of the world's market for mousetraps. The trap snaps shut at 38,000 of a second.

CEREMONIAL BELL, PATENT NO. 110,875
John W. Smith, Keokuk, Iowa, January 10, 1871

This bell was designed for use during certain ceremonies, which were conducted by various secret and other societies, that required the tones of a bell gong. One or more bells or gongs are arranged in a box in such a manner that they can be conveniently struck by means of a gavel and not be exposed to view. The perforations or openings through the sides and top of the box are covered by a suitable cloth. In front of the bell is a sounding board which is perforated to conduct the sound in the direction of the openings through the vertical sides of the case. The gavel has a long handle attached to a cylindrical head which is covered with a soft velvet material.

WINDOW CLEANING STEP CHAIR, PATENT NO. 206,936
Ana Dormitzer, New York, New York, August 13, 1878

Cleaning the outside of windows on upper stories could be difficult and dangerous. This device provided a throughly safe as well as a convenient method for cleaning windows. The chair could temporarily be attached to any window, and support a person safely while cleaning the outside of windows. After use it could compactly fold together, requiring little room for storage. The chair is put into position on the window sill and held in place by a brace attached to the outside of the wall with thumb screws.

EARTH SCRAPER, PATENT NO. 204,142
James H. Edmondson,
Valparaiso, Indiana, May 28, 1878

The object of this invention is to furnish an improved sulky road scraper, which shall be so constructed that it may be easily operated by the driver from his seat, to load and unload it, which when loaded, may be swung beneath the axle and carried to any desired distance. It shall be simple in construction, convenient in use, and effective in operation.

BALLOT BOX, PATENT NO. 226,689
Joseph Welch, Delphos, Ohio, April 20, 1880

This ballot box is intended especially for the use of secret societies, and is constructed for the purpose of preventing those around from seeing how each voter votes. An open space is in the top of the box, in which the white and black marbles are kept ready for the voter, and which opening is so constructed that a person can see down into it from the top of the box and thrust his hand into it from one side. The marbles then slide down a track and are collected in a drawer with a hinged top. When the voting is completed, the drawer is opened and the marbles are displayed. A white marble constitutes a vote in support and a black marble signifies opposition. The term "blackballed" is derived from this type of voting. The word ballot comes from the french word ballotte or "little ball."

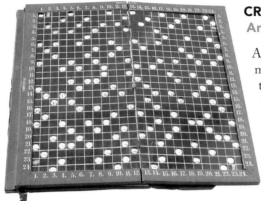

CRYPTOGRAPHY, PATENT NO. 166,761
Anthony Luis Flamm, Paris, France, August 17, 1875

Anthony Flamm, an Ex-Major of the Ottoman Army, invented a new and improved method of sending secret correspondence. The cryptograph consists of a tablet forming twenty four each vertical and horizontal columns which produces 576 squares. The plate is pierced with 144 holes and by placing the plate over the tablet, turning it each time 1/4 round, either to the left or right as agreed upon by the parties involved, the 576 squares will be used. The distribution of the holes in the plates may be varied to an almost unlimited extent thus allowing the encrypted message to be read by only the individuals with the same hole configuration as the message was sent on. The holes are filled in with letters that make up the correspondence and the plate is rotated in a configuration that is only known by the sender and receiver.

ELECTRO MAGNETIC ENGINE, PATENT NO. 122,944
Claude Victor Gaume, Williamsburg, New York, January 23, 1872

This motor was used for operating sewing machines and other light machinery. It was constructed to be free from "pull back" or retardation which was a great objection to such engines as usually constructed. When the motor is connected with a battery, the current of electricity passes through the frame, shaft, and arms to the plates. As the wheels come upon the small plates, the circuit is closed, and the current passes on through the wheels, arms, and wires to the coils, and through the wires back to the battery. When the circuit is closed, the magnets attract the armatures, and when the circuit is broken, the magnets cease to attract the armatures.

PREPARING FISH FOR MARKET, PATENT NO. 161,596
David W. Davis, Samuel H. Davis, Detroit, Michigan, April 6, 1875

This innovation reduced losses of frozen fish due to breakage while being handled. In handling fish, when frozen either singly or in small parcels, they are liable to be broken, resulting in the price being very reduced. In order to obviate such loss, fish were frozen into cakes using pans that would pack neatly into a cask for transport. The cask could then be opened and a single frozen cake removed without disturbing the other cakes, reducing handling and damage to the fish.

PNEUMATIC DESPATCH CARRIER PATENT NO. 648,853
James T. Cowley, Lowell, Massachusetts, May 1, 1900,
Assignor to the Lamson Consolidated Store Service Co.

The carrier is constructed so that it is impossible to insert the carrier into the pneumatic despatch tube until the cover is closed and locked to the body of the carrier. Once the carrier is inserted into the tube, the cover cannot become unlocked and opened while the carrier is in transit. The carrier is made of sheet metal or other suitable material and is provided at one end with a fixed head of felt. The opposite end is provided with a movable cover that is lockable to the carrier. The Lamson Company today is the premiere supplier of pneumatic tube systems throughout the world.

UNIVERSAL JOINT, PATENT NO. 197,541
Phineas Burgess, New York, New York, November 27, 1877

Designed especially for propeller shafts, this universal joint was equally applicable to other heavy shafts where great strength was required. It was constructed as to very greatly increase the strength, without materially increasing the amount of metal used in its being produced. The construction of the lugs added great strength to the joint and the bolts, being supported upon both by the outer and inner sides of the ring, would receive the strain squarely, without any tendency to bind, so that the movement of the joint would be easy and uniform.

BLACKBOARD AND MAP CASE, PATENT NO. 40,035
William C. Herider, Miami Town, Ohio, September 22, 1863

A rectangular case is employed to make a combination of a blackboard and map case for use in schools. The case is constructed with a hinged door on its front side taking up half the length of the front of the case. The whole of the front of the case is painted black so as to serve as a blackboard. The interior of the case at its top and bottom is provided with guides or grooves in which frames are fitted and on which maps are secured. The case may be designed to accommodate up to ten maps.

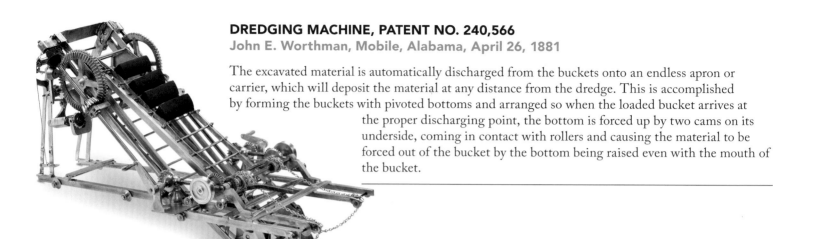

DREDGING MACHINE, PATENT NO. 240,566
John E. Worthman, Mobile, Alabama, April 26, 1881

The excavated material is automatically discharged from the buckets onto an endless apron or carrier, which will deposit the material at any distance from the dredge. This is accomplished by forming the buckets with pivoted bottoms and arranged so when the loaded bucket arrives at the proper discharging point, the bottom is forced up by two cams on its underside, coming in contact with rollers and causing the material to be forced out of the bucket by the bottom being raised even with the mouth of the bucket.

RUDDER INDICATING APPARATUS FOR VESSELS, PATENT NO. 174,186
Justus A. Briebach, Clapton, England, February 29, 1876

For the prevention of collisions at sea, this device is connected to, and operated automatically by, the rudder or helm, for the purpose of showing at a distance, the position of the rudder which will indicate the direction in which the vessel is steering. The apparatus has two operating warning systems. For daytime use, two different colored flags are used to indicate the direction of the rudder and for nighttime, a lighted lamp is used with red and green lenses to indicate the vessel's direction.

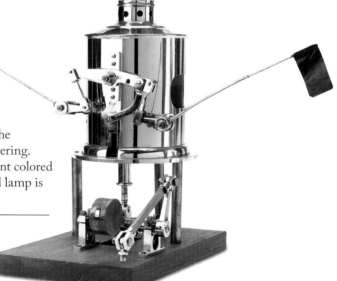

CHAPTER 19

Personal

"To invent you need a good imagination and a pile of junk."

— Thomas A. Edison

They say "time and tide wait for no man." The march of time is relentless, and eventually, the hourglass sand runs out on all of us. In this chapter, we look at a few critical inventions related to the keeping of time, some designed to ease us out of this world in style when the clock runs out.

MARTIN HALE CRANE (1821-1886)

In 1848, Almond Fisk patented an "air-tight coffin of cast or raised metal." The coffin resembled an Egyptian sarcophagus with sculpted arms and a glass window for viewing the face of the deceased. Fisk suggested removing all air from the coffin in order to prevent putrefaction. Seeing an opportunity in Fisk's coffin design, Martin H. Crane and his partner Abel D. Breed of Cincinnati, OH, purchased the Fisk Metallic Burial Case Company in 1853. The partners had incorporated Crane, Breed & Company in 1850 and ran the company until around 1880. In 1855, Martin Crane patented a "metallic coffin" that modified Fisk's original design, eliminating the mummy shape and simplifying the decorations to make it easier to mass produce. Fisk and Crane's Patent Metallic Burial Case claimed to preserve the body, protect it against water and vermin, safeguard against contagion, and facilitate relocation of the remains. Fisk and Crane cases came lined with muslin or satin, padded with cotton, and were sold with a packing crate. The Crane metallic burial case was rectangular in shape, an important step in the evolution of the modern coffin or casket design. The first to be called a "casket," Crane's burial case permitted viewing of the entire body. The lower-priced casket opened up new markets for the firm, especially in the Southern U. S. where they established a satellite sales office and showroom in New Orleans, LA.

The 1860-61 Cincinnati City Directory listed the firm as Crane, Breed & Co., manufacturers of stoves and hollow ware, also patent metallic burial cases and caskets. In the 1870 census, Crane was listed as a foundry manager, and in the 1880 census as a coffin manufacturer.

Martin Crane began experimenting with rolled sheet iron in the mid 1860s as a less expensive alternative to cast iron,

Figure 1: Advertisement Crane & Breed Co.

and by the end of the decade, he had perfected the industry's first sheet metal casket. The company was considered to be one of America's premier builders of horse drawn hearses, and introduced the first commercially available automobile hearse in 1909. President Abraham Lincoln was interred in a Crane coffin. (Figure 1)

METALLIC BURIAL CASE, PATENT NO. 64,496
Martin H. Crane, Cincinnati, Ohio,
Assignor to Crane, Breed & Co., May 7, 1867

This burial case combined strength, lightness, and economy of labor and material with a handsome appearance. The case was composed of sheet zinc metal carefully soldered at the joints to form a tight vessel. The upper and lower edges of the case were stiffened and ornamented by sills and rails composed of strips of sheet metal stretched over wooden strips that were firmly soldered to the exterior of the case. The bottom and lid were similarly stiffened.

WILLARD LEGRAND BUNDY (1845-1907)

Willard L. Bundy was a jeweler and inventor who invented an early version of the time recorder, a mechanical device which recorded when workers clocked in and out of their jobs. In Bundy's system, each employee had a numbered key which they inserted into the recorder to clock in or out. Inserting the key printed the key number and the time on paper tape. The system was considered more efficient than human time keepers.

Willard and Harlow Bundy founded the Bundy Manufacturing Company in 1889 to produce the time clock that Willard had invented. The brothers began with eight employees and $150,000. Willard Bundy was superintendent and continued to invent, while Harlow, an attorney, managed the company and promoted the product. By 1898, nearly 9,000 Bundy time recorders were on the market. The company advertised the devices as being guaranteed to solve "vexatious questions of recording employee time." Early Bundy clocks used various advertising slogans such as "Your time clock is your purse. Are the strings drawn taut?", "Time is your chief stock in trade," and "Place value on every minute with our machines." The federal government used Bundy machines in post offices, saving the government millions of dollars in time keeping labor. Bundy's eight-foot-tall clock was exhibited at the 1876 Centennial Exhibition.

In a dispute over patents, Willard left his company in 1902 and started Bundy Time Recording Company.

Bundy's original concept of a time clock is still used in businesses throughout the world today. Even though he left the initial company, Bundy's role as founder and inventor contributed to the success of Bundy Manufacturing Company. Over the years, the company went through a series of mergers, eventually becoming part of International Business Machines (IBM). It is currently part of the Simplex Time Recorder Company. (Figure 2)

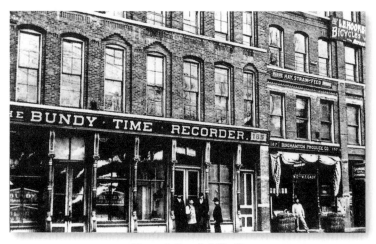

Figure 2: Bundy Mfg. Co.

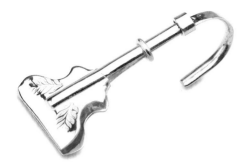

NAPKIN HOLDER, PATENT NO. 217,264
Willard L. Bundy, Auburn, New York, July 8, 1879

Previous napkin holders had hinged clamping jaws that were forced together by independent springs, which were expensive to produce and prone to breaking, or that poked holes in the napkins. Bundy's napkin holder used two spring arms with an enlarged clamp at one end and a hook at the other. The ring slid along the arms, and the clasp united the arms to hold the napkin in place.

ADRIEN PHILIPPE (1815-1894)

Adrien Philippe (Figure 3) was the son of a watchmaker who followed in his father's footsteps, but left home at the age of 18 to learn from other master watchmakers. In 1842, Philippe invented a winding and setting mechanism for watches that used a crown rather than a key. Before getting his invention patented in the U.S., Philippe patented it in France. On April 22, 1845, he obtained French Patent No. 1317 for the Keyless Watch. This keyless stem winding and setting system (pull out to set, push in to wind) revolutionized watch making.

His patented invention earned him a Gold Medal at the French Industrial Exposition of 1844. There, he met Antoine Norbert de Patek, who was so impressed with Philippe's accomplishments that he offered him a position in his company. Philippe would become head watchmaker at Patek & Co, in Geneva under an agreement that entitled him to one third of all company's profits. By 1851, Philippe was a full partner and the company's name was changed to Patek Philippe & Co. At London's Crystal Palace Exhibition in 1851, Queen Victoria purchased Patek Philippe watches for herself and for Prince Albert. Patek and Philippe's partnership capitalized on their individual

Figure 3: Portrait of Adrien Philippe

talents. Patek was a talented salesman as well as an avid traveler, who traveled around the world to promote and market his firm's watches. Philippe focused his efforts on overseeing the technical direction of the firm, as well as overseeing its day-to-day operation.

In 1863, Philippe published a book in Geneva and Paris on the workings of pocket watches entitled, *Les Montres San Clef* (Watches with Keyless Works).

Patek Philippe pioneered many new watch features. As early as the Paris Exposition of 1867, Philippe's watches featured functions that were to become the standard for the complicated mechanical watches of the 20th century. Philippe introduced technical improvements such as the free mainspring and the sweep second hand, in addition to improvements to regulators, chronographs, and perpetual calendar mechanisms.

His watch system would lead to the development of the wristwatch, which women first started wearing in the late 1800s. During World War I (1914-1918), wristwatches became increasingly popular as soldiers realized that they were much more convenient than pocket watches. Philippe's self-contained, keyless watch evolved into today's waterproof wristwatches. Philippe was named a Knight of the Legion of Honor in 1890 in recognition of services rendered to the country of France.

Patek Philippe watches over the years have won more than 500 international awards and are considered the "Rolls Royce of watches." Their watches today, as in the time of Philippe and Patek, are still made entirely by hand and have now dominated the Swiss watch industry for over 160 years.

SPRINGS AND BARRELS FOR TIME KEEPERS, PATENT NO. 43,464
Adrien Philippe, Geneva, Switzerland, July 5, 1864

This invention obviated the need for the "fixed stop." By dispensing with the hooking of the spring to the barrel, the design allows the spring to turn indefinitely in the barrel without injury to the watch and all chance of accident is avoided. When the spring is fully wound, "a projection falls into one of the channels and produces a slight click, sensible to the ear and to the touch. Thus the person winding is advised that the watch is fully wound."

TRAVELING TRUNK, PATENT NO. 67,905
Louis Ransom, Lansingburg, New York, August 20, 1867

This cylindrical trunk had iron staves that served as wheels; it could easily be moved around by rolling. An amazing fact: it took more than 100 years for another inventor to figure out that putting wheels on luggage would make handling luggage much easier. In 1970 Bernard Sadow was issued Patent No. 3,653,474 for "Rolling Luggage." One fine example of "Yankee Ingenuity" by Ransom in 1867.

LAP BOARD, PATENT NO. 135,040
Sara Mahan, Cleveland, Ohio, January 21, 1873

This lap desk's legs took some of the weight of the board off the person's lap. The legs could hold the board level, so articles on the board did not slip off. The third leg stabilized the board when it was set aside.

BOOT JACK AND BURGLAR ALARM COMBINED, PATENT NO. 19,844
F. C. Coffin, Newark, New Jersey, April 6, 1858

This device was a rare combination, serving the purpose of a Boot Jack and a Burglar Alarm. No other patent was ever issued for this combination of a device by the U.S. Patent Office. As a boot jack, the movable jaws firmly grabbed the boot heel and facilitated removing the boot. As an alarm, the boot was left in the device, the spring arm was activated, and the whole device placed on the floor against the inside of the door being protected. Opening the door triggered the sliding arm, causing the boot to fly up and hit the door with a loud noise, thus alerting the occupant to the intruder. The implement was compact and portable, making it very useful in traveling.

COMBINED LUNCH BOX AND SHOPPING BAG, PATENT NO. 144,385
Charles C. Cobleigh, Brighton, Massachusetts, November 11, 1873

This unique combination provided a simple, compact and convenient lunch box and shopping bag for those who wanted to have a meal and then go shopping. The lunch box consisted of removable galvanized tin food containers and a canteen. To use it as a shopping bag, the food receptacles would be removed, leaving plenty of room in the bag for parcels. The bag was made of leather lined with tin or nickel plate.

CONSTRUCTION OF TRUNKS AND VALISES, PATENT NO. 520
Mathias Steiner, New York, New York, December 20, 1837

In this trunk, a metal frame was riveted around the entire opening of the trunk, both inside and out. The framework inside the trunk rendered it tight and waterproof. The trunk or valise could be secured by inserting a metal rod through the metal frames and locked with a special key. For further security, the trunk could have an additional hasp and padlock.

MIRROR ADJUSTERS, PATENT NO. 178,234
Sarah M. Clark, Milwaukee, Wisconsin, June 6, 1876

This system used pivots to position a mirror. A metallic toothed semicircular plate is attached to the mirror. A pawl operates automatically in the tooth edge of the metallic plate as the mirror is inclined forward or backward and holds the mirror at any given angle desired.

CALENDAR, PATENT NO. 254,015
Thomas H. Hovenden, Ingersoll, Ontario, Canada, February 21, 1882

This perpetual calendar indicated the day of the week, the month, and the day of the month. A single knob could set either the name or number of the day or the name of the month. The calendar was designed to keep it within a reasonable size, with simple moving mechanisms that would produce an effective and practical calendar within reasonable limits as to cost.

HAT HOLDER, PATENT NO. 80,374
Zera Waters, Bloomington, Illinois, July 28, 1868

This invention addressed the problem of people taking the wrong hats from racks, saving those responsible for the loss of hats a great deal of expense in replacing the hats. It was particularly designed for hotels, steamboats and public halls. Only the key holder could remove the hat, thus preventing "the theft of the hat by choice or by mistake."

INTEREST CALCULATOR, PATENT NO. 271,949
Marshall Todd, Danville, Indiana, February 6, 1883

This device could calculate interest on any desired sum of money for any desired number of days, months or years. The interest computer was contained in a covered box with apertures for showing the digits of the calculation results. Inside, the box contained a series of marked subdivisions with apertures of sliding cards or charts arranged one above the other, so the numerals indicated the interest values.

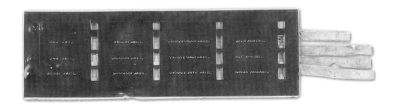

SUSPENDERS, PATENT NO. 211,999
Harry M. Heineman, San Francisco, California, February 4, 1879

Ordinary elastic suspenders had been made of leather or other textile that typically wore out or broke where they attached at the button holes. These suspenders used braided or twisted wire for the connecting pieces of the straps. The center of the strap was wound around a ring attached to a hook, and the two ends led down to the buttons.

VEGETABLE FIBER TO IMITATE HAIR, PATENT NO. 89,855
Andre Couturier Trinidad, Cuba, May 11, 1869

The fibers from the pita or leaf of the Corojo Palm, which grows in the West Indian Islands, had "superior qualities and unequaled strength relative to other animal or vegetable fibers that are being used in manufacturing." The fibers being of great length can be spun into the very strongest thread. These threads, which appear similar to human hair, when colored can be used as imitation hair.

BURIAL CASKET, PATENT NO. 202,440
John Homrighous, Royalton, Ohio, April 16, 1878

A casket had a rectangular shape, with all the sides being straight. Constructing a burial case that could be adjusted to various lengths reduced the number of caskets that the undertaker would normally have to keep in stock. The foot section of the casket is recessed from the head section allowing the case to be extended to various length sizes. The use of this casket reduced the nineteen different size caskets manufactured to only needing six adjustable caskets.

MEASURING SCALES FOR COFFINS, PATENT NO. 58,262
James W. Hyde, Lewiston, Illinois, September 25, 1866

The shape of a coffin is narrow at the head, wider in the middle and narrow again in the bottom or foot section. The measuring scales had graduations that allowed the coffin maker to lay out the configuration for the top and bottom portions for the wood pieces needed to construct the coffin. The use of these measuring scales help reduced the amount of wasted stock in making a coffin.

BED BOTTOM AND FIRE ESCAPE, PATENT NO. 211,275
Frederick Swinden, Alfred Buxton, Naugatuck, Connecticut, January 7, 1878

In case of a fire, this bed bottom could be easily converted into a convenient fire escape. The bed bottom's five individual sections of springs (the ladders) rested on a cross bar on the frame of the bed bottom. In case of a fire, the cross bar was removed so the ladders could be removed and extended. The hooks allowed several sections to be connected together. Finally, the ladder was let out through a window and secured by two chains inside the casing of the window.

DESIGN FOR A COLLAR BOX, DESIGN PATENT NO. 4,446
Albert Aronson, New York, New York, November 1, 1870

On August 29, 1842, designs were made patentable. A separate numbering system was initiated starting with Design Patent No. 1, issued to George Bruce for Printing Types. Collar boxes were popular during the 1800s to store the collar portion of a man's shirt in a fancy box to keep it from getting dirty when not in use. This collar box represented a log cabin with a door and windows, with a sign over the door giving the name of the collar that belonged in the box.

CHAPTER 20

Healthcare

*"The only way to keep your health is to eat what you don't want,
drink what you don't like, and do what you'd druther not"*

— Mark Twain

Medicine in early America was a combination of home-made folk remedies and more scientifically-based practices carried out by doctors, many of whom lacked proper medical training and credentials. During the early Colonial and early post-Colonial periods, physicians learned their trade primarily through apprenticeship. It wasn't until 1751 that the first hospital opened, and it wasn't until 1765 that the University of Pennsylvania opened the first medical college in Colonial America.

In the early 1800s, physicians began to fully understand the concept of germs and the fact that social and environmental conditions led to the spread of certain diseases, especially amongst urban populations. Industrialization brought more people to cities in search of work, and this often led to overcrowding, poor sanitation, and devastating epidemics within these tightly packed populations. Cholera, diphtheria, tuberculosis, and yellow fever, along with concerns over sanitation and hygiene, led governments to create departments of health.

In the mid 19th century, France, Germany, and Britain were considered to be the world's major medical centers; the U.S. was thought of as nothing more than a backwater. Reputable American medical schools were still in their infancy, and the majority of American physicians of the time were poorly trained.

Because of the lack of medical knowledge and effective treatments, unorthodox (and "quack") medicine flourished. Tobacco, herb, and root concoctions were used to cure ailments like toothaches and gout. Physicians prescribed opium-based drugs, and morphine was used as a common painkiller. Patent medicines were coming to market with no government regulation or restrictions on sales. These quack

cures could even prove deadly, as there was no regulation on their ingredients. By the mid-1800s, the manufacture of these products had become big business in America. Often high in alcoholic content, many were spiked with morphine, opium, or cocaine. Originally, patent medicines referred to drugs whose formulations had been granted government protections for exclusivity. In truth, most 19th century patent medicines were not officially patented, and most used very similar ingredients.

The most common surgery for infections in the 18th century was amputation. Sterilization of operating tools was not practiced, doctors being unaware that unsanitized tools actually created more infection and disease. During surgery, patients were operated on while still awake. Anesthesia was finally introduced in the early 1840s. It's perhaps not surprising that Americans born in 1876 had a life expectancy of little more than forty years, with the scientific literacy in the field of medicine still in its youth and medical care in America often inadequate, if available at all.

The 19th century brought great advances in our understanding of health and the science of medicine. Among other things, an acceptance of the link between environment and health was finally widely acknowledged. And there was increased understanding of germ theory and the power of social reform as a way of improving the public health. Two giants who championed these advances were Louis Pasteur (1822-1895) and Joseph Lister (1827-1912).

Louis Pasteur discovered that microbes were to blame for souring alcohol and milk. He came up with a process of pasteurization in which bacteria are destroyed by boiling beverages and then cooling them down. He completed his first successful tests of the process in 1862. His work in germ theory also led him to create vaccinations for anthrax and rabies. Pasteur's first vaccine discovery was in 1879, for a disease called chicken cholera. After accidentally exposing chickens to a form of the culture, he found that they became resistant to the actual virus. As a result of this discovery, Pasteur went on to develop vaccinations for anthrax, cholera, TB, and smallpox.

Joseph Lister was the surgeon who introduced principles of cleanliness that changed surgical practices of the late 1800s. Prior to the change, a patient could undergo a medical procedure successfully, only to die from a postoperative infection. Lister read Pasteur's work on microorganisms and experimented with exposing the wound to chemicals. He used dressings soaked with carbolic acid to cover the wound and found the rate of infection to be greatly reduced. He went on to experiment with hand washing, sterilizing instruments, and spraying carbolic in the operating room to limit the chance of infection. These practices succeeded in greatly reducing infection rates. Lister's experiences with sterile procedures in surgery, supported by the research of Pasteur and others, had a significant impact on the practice of medicine in America. For his efforts and discoveries, Lister is now known as the "father of antiseptic surgery."

Many products, with names still familiar to us, were invented in the late 1800s. Listerine, named after Lister, was developed in 1879 by Jordan W. Lambert, cofounder of the Warner (Warner-Lambert) pharmaceutical company, as an antiseptic and disinfectant. Felix Hoffmann synthesized acetylsalicylic acid, and in 1899, the Bayer Company, where he was employed, began marketing what was to become the most popular of all patent medicines, aspirin. In 1880,

Charles Henry Phillips developed an antacid, a white suspension of magnesium hydroxide in water, which he called "Milk of Magnesia." Lunsford Richardson invented a cold remedy in the 1890s that he called Richardson's Croup and Pneumonia Cure Salve. He later renamed the product Vick's Salve, and later still it became Vick's VapoRub.

Before the 19th century, dental treatment had been torture carried out with the crudest of instruments. Extraction often ended with the fracture of a patient's jaw. Artificial teeth, made from animal teeth or tusks, or from human teeth, caused halitosis and were subject to deterioration and discoloration. Ligatures for securing replacement teeth in the mouth were crude and uncomfortable.

The 1800s would see many great innovations in dentistry. In 1825, Samuel W. Stockton began experimenting with casting porcelain and soon began making teeth. His nephew, Samuel S. White, indentured to his uncle at the age of 14, became a leader in the design and manufacture of dental equipment. White developed artificial teeth, dental instruments, dental chairs, and a dental engine that could be powered by electricity. White's artificial teeth included many innovations such as translucency without loss of strength, increased resistance to temperature variations, and improved means of attachment to denture bases. A major factor in the improvements was the availability of hard rubber. Charles Goodyear, Jr., in 1855, patented a dental plate made of rubber "vulcanized" by an alloying and heating process. This flexible but durable material was the best yet available for denture bases, but the Goodyear patent monopoly covering vulcanite kept costs high. White spent much time and energy in a patent war with Goodyear, which White ultimately won.

ERNST SCHERING (1824-1889)

Ernst Christian Friedrich Schering (Figure 1) created the Schering Corporation, and in Berlin in 1851, he opened a pharmacy that he named Gruene Apotheke (Green Pharmacy). Twenty years later, he incorporated the business as Chemische Fabrik aug Actien. In 1889, Schering died at the age of 65, one year before his company began marketing its first specialty product, a medication for gout. In 1901, the company's operations expanded into electroplating. These operations would soon expand to include production of complete electroplating equipment, chemical compounds, and machinery for printed circuit manufacturing.

At the beginning of the 20th century, the German company also expanded into industrial and laboratory chemicals. In the

Figure 1: Portrait of Ernst Schering

1920s, agrochemicals were added to Chemische Fabrik's product line, and by the end of the decade, they had entered into the area of female hormonal products. In 1937, the company merged with Oberkoks, a mining and chemical company, becoming Schering AG. Schering's operations in the U. S. were adversely affected by the two World Wars. During the first World War, the company's U.S. subsidiary was dissolved. After being reestablished, during the second World War in 1929, it was seized by the U.S. Alien Property Custodian Office. The company split into two companies as a result of the seizures, Schering AG and Schering Plough. In 2009, Merck & Co. merged with Schering Plough under the name Merck & Co.

PURIFYING SALICYLIC ACID BY DIALYSIS, PATENT NO. 196,254
Ernst Schering, Berlin, Germany, October 16, 1877

In this process, carbonate of sodium is mixed with carbonic acid gas in a closed retort to produce salicylate of soda. The salicylate is dissolved in water and added to a solution of mineral acid, which precipitates out salicylic acid. The salicylic acid is again dissolved in water and heated with zinc dust, to whiten the solution. This solution and alcohol are filtered and percolated through the membrane sides of a tank which filters out foreign matters in the acid. The filtered salicylic acid is cooled in the presence of sulfuric acid to precipitate the salicylic acid in the form of bright needles or crystals.

BENJAMIN JEWETT (1853-1928)

The Civil War changed the fields of military surgery and orthopedics. The powerful, destructive, and more accurate weapons of this conflict did more damage to the human body than ever before, resulting in not only more injuries, but injuries of a more critical nature. The medical profession was unable to keep up with the gruesome advances in deadly weapons. Wartime surgeons had little understanding of sterilization and post-surgical infection, so amputation was commonplace. Wounded soldiers taken from the battlefield were given whiskey, or when available, opium. If surgery was to be performed, soldiers were given chloroform or ether. If an amputee did not die from blood loss, he was likely to die from post-surgical infection and gangrene. Dressings were often reused and rarely cleaned.

Beginning in 1862, the Federal government provided money to Union amputees with which to purchase artificial limbs. Amputees were given fifty dollars for an arm and seventy-five dollars for a leg. In the state of North Carolina, a resolution was passed in 1866 to begin a program similar to the Federal one, to supply legs to Confederate veterans. North Carolina also contracted with Benjamin Jewett to supply the artificial limbs.

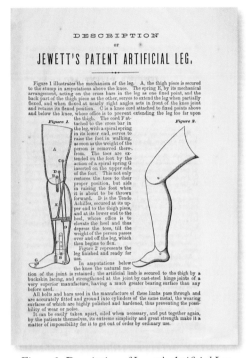

Figure 2: Description of Jewett's Artificial Leg

Jewett's Patent Leg Company provided two proposals to North Carolina, the first offered exclusive rights to his artificial leg, in the form of two patents for it, for a sum of twelve thousand dollars. The second option had Jewett setting up shop in North Carolina and charging the state seventy-five dollars a leg, with an advance of five thousand dollars. Jewett also requested that the state provide a suitable building for manufacturing the limbs. The state agreed to the second option. Like the Federal artificial limb program, North Carolina's plan ensured that amputees did not incur out-of-pocket expenses during their stay. Jewett's Patent Leg was praised in a Civil War report on artificial legs for its "mechanical perfection and simplicity of arrangement." It was also noted for the ease with which a wearer could take it apart for minor repairs. Jewett's Patent Leg Company in Raleigh stayed open until June 18, 1867. At this point, having served the soldiers that had returned from the war, the shop no longer had enough work to continue on-site manufacturing.(Figure 2)

ARTIFICIAL LEG, PATENT NO. 35,686
Benjamin B. Jewett, Salem, Massachusetts, June 24, 1862

Past artificial leg sockets were difficult to adapt to the stump of the leg and even more difficult to readjust to the limb after a period of usage. This invention produced a rigid but comfortable leg socket that could be readily modified to fit any stump. The length of the leg could vary, and any part of the leg could be modified or replaced without altering the rest.

INVALID BEDSTEAD, PATENT NO. 19,254
George Miller, Fremont, Ohio, February 2, 1858

This invalid bed could be converted into an easy chair. The lower portion of the frame of the bed was connected by cords to a windlass, which was turned to draw the frame upward. Any desired inclination could be given the back by stopping the rotation of the windlass. This particular patent model is very unusual because it is the only model in the collection where the model maker used tiger maple wood.

PAPER BOARD FOR DRESSING WOUNDS, PATENT NO. 257,723
Dr. Paul Koch, Nuffen, Wurtemberg, Germany, May 9, 1882

This board was used to support a wounded or injured limb. The pulp for the board was hammered to make it soft, then steeped in "an alcoholic solution of shellac, resin, and turpentine," then rolled out after it dried. To apply, it was first steeped in warm water and then molded by hand to conform to the shape of the injured area. This invention was also patented in Germany, Belgium, France, England, and Hungary.

ARTIFICIAL PALATE, PATENT NO. 164,591
Dr. Jacob Peyer, Berne, Switzerland, June 15, 1875

This artificial dental palate was intended to replace the total or partial loss of the natural palate in a patient. It consisted of an oval plate provided with a sheath, in which a flat plate or tongue attached to the palate, was adapted to fit and move with a slight longitudinal motion. Two wings are pivoted on the plate and pressed outward by means of a spiral spring to adapt the valve movement to the muscles. The metal used with the rubber in making this apparatus was preferably gold or platinum.

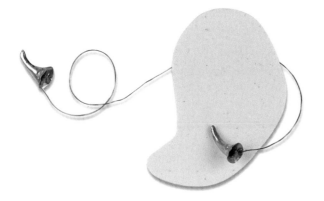

EAR TRUMPET, PATENT NO. 78,493
Dr, Thomas H. Stilwell, New York, New York, June 2, 1868

This innovation consisted of two small trumpet-shaped tubes that were inserted into the ear canal so they barely projected outside of the ear. The tubes were connected by a copper wire, with an electrical current running between the tubes "to invigorate paralyzed or weakened auditory nerves." The wire would also "increase the vibrations of the air, thereby making sounds appear louder to the patient."

ELASTIC WOVEN FABRIC, PATENT NO. 484,977
Max Picot, Paris, France, October 25, 1892

The fabric was formed entirely of cotton, wool, linen, or silk or a combination of the materials. The fabric was longitudinally, elastic making it especially advantageous for use in surgical bindings, being devoid of seams and of any length necessary. The elasticity of the bandage causes it to remain firm when it's used as a binding and results in a slight pressure on any part of the body. It freely allows movement of the bound up part without slipping off or causing pain.

FRACTURE BOX, PATENT NO. 163,829
Christian Westerkamp, Cincinnati, Ohio, May 25, 1875

This hinged fracture box was for treating broken legs and limbs. The box consists of two bottom plates joined together with hinges. The short end of the bottom plate joined to the lower foot plate and is intended to enclose the lower part of the thigh. The lower part of the broken leg is inserted in the long part of the box, with the sole of the foot placed against the foot board. The side boards opened to allow the leg to be bandaged, and the square holes in the side boards were for straps to tie the leg firmly in the box after the bandage had been placed upon it.

PILL MACHINE, PATENT NO. 243,848
Pierre Cauhape, New York, New York, July 5, 1881

This machine manufactured pills by using a set of two-part hinged molds. A comb bar is adapted with pins for removing the pills from the mold and then dipped them into a bath of glycerine or other substance with which they are to be coated. The pills were dipped a second time to cover any portion left uncoated by the first operation and closed the holes left by the pins.

MACHINE FOR MAKING PILLS, LOZENGES, &c., PATENT NO. 189,005
Thomas J. Young, Philadelphia, Pennsylvania, March 27, 1877

To shape the pills, dry powders were compressed between a reciprocating plunger or an upper die and a lower die, which remained stationary during compression and then ejected the finished pill. A lozenge is a small, sometimes medicated tablet that is slowly dissolved in the throat to relieve soreness and coughing.

PORTABLE COMMODE, PATENT NO. 223,574
William S.G. Baker, Baltimore County, Maryland, January 13, 1880

This invention provided a portable seat for the infirm. The commode placed a seat with an opening above a suitable vessel. It was made to be readily and quickly set up for use or folded closely together making it easy to transport.

LATHES FOR DENTAL SURGERY, PATENT NO. 220,384

George Horatio Jones, 57 Great Russel Street,
Bloomsbury, County of Middlesex, England, October 7, 1879

The purpose of this invention was to save the small but valuable particles of precious metals, such as gold and platinum, that were wasted in the operation of grinding when using lathes for dental purposes. When the driving wheel of the lathe is set in motion, water is pumped up through a tube and passes over the grinding wheel. The water is discharged into a tank, that carries with it all the minute particles of precious metals which become detached in the operation of grinding. When a sufficient accumulation of particles had been collected, they were removed and submitted to the ordinary process of refining for reutilization.

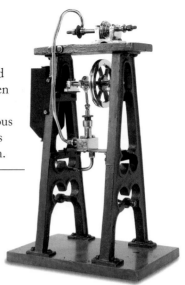

DENTAL HAND PIECE, PATENT NO. 233,709

Eli T. Starr, Philadelphia, Pennsylvania, October 26, 1880

A dental hand piece is an instrument that holds various disks, cups or burs that a dentist uses during various dental procedures. This dental hand piece was designed so it acted steadily and positively, without rattling or vibration. It also incorporated "a multitude of innovative inner workings" such as an improved lock and release for the cutting tool and new collars and bearings. Two patents were granted to Starr in 1879 for the treadle pedal style engine that delivered power to this hand piece.

INVALID'S TABLE, PATENT NO. 23,068

Jonathan M. Allen, Worchester, Massachusetts, March 1, 1859

This invalid table was attached to a tubular column that ran between the floor and the ceiling. The table height was adjustable, and it could swing around from side to side. The column was also useful in assisting a patient to rise from the bed. It could be firmly and securely supported in any part of a room without occupying a considerable amount of space.

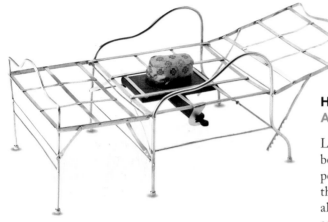

HOSPITAL BEDSTEAD, PATENT NO. 77,193
Anthony Iske, Lancaster, Pennsylvania, April 28, 1868

Light castings and wrought iron were used in the construction of this hospital bedstead. It differed in the general character of hospital beds during this period of time because the bed was made in three distinct sections, the head, the foot, and the body section. The bed contained a built-in commode and also was capable of regulating or raising and lowering the patient's head and shoulders. The inventor stated that "the present arrangement is highly approved of in several hospitals to which samples have been given on trial and it is declared to possess all the facilities that could be desired."

ARTIFICIAL LEG, PATENT NO. 37,282
Theodore F. Engelbrecht, Reinhold Boeklen, William Staehlen, Brooklyn, New York, January 6, 1863

This artificial leg could be adapted to the length of the natural limb and conformation of the foot of the intending wearer. The necessity of making a limb to suit each particular case is to a great extend obviated, and in consequence the cost of manufacture is considerably reduced. The novel construction of the foot portion permitted the ankle joint to approximate nearer to the natural joint than other artificial legs constructed during this period. The leg was constructed of corrugated metal plate and preferably covered with leather.

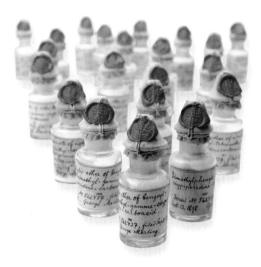

COMPOUND OF GAMMA OXYPIPERIDIN CARBOACIDS, PATENT NO. 591,483

Georg Merling, Berlin, Germany, September 13, 1895

This patent for certain new and useful improvements in The Process of Preparing Nitryls of Gamma Oxypiperidin Carboacids and Derivatives is rather unusual. An actual patent model was not submitted with the application, but instead, twenty two specimens of different derivatives of the compound were submitted in six wooden boxes. After researching this patent for the past 14 years, the only resulting information obtained was that two Polish resorts at one time offered Carboacids Baths but it has been impossible to determine what value these baths were supposed to provide. Another possibility is that the derivatives may have some similar characteristics to our modern-day psychedelic compounds.

CLUB FOOT SHOES, PATENT NO. 167,867

William Autenrieth, Cincinnati, Ohio, September 21, 1875

Club foot is a congenital birth deformity involving one or both feet where the affected foot appears to have been rotated internally at the ankle. This shoe attachment was intended to bring club and twisted feet back to their proper position by making a gradual adjustment. The shoe is put on the patient and the screw on the foot plate is turned gradually with the key a little every day, until the foot is brought to a normal position.

CHAPTER 21

Music

"Music is a world within itself, with a language we all understand."

— Stevie Wonder

In many ways, the roots of American music can be traced back to the African slaves being brought to America. The first well-recognized form of African-American music was spirituals which originated in the Southern U.S. among slave populations. Spirituals were religious songs that told a story and expressed deep emotions. Spirituals sung by African slaves often spoke of their hope, faith, and their longing for freedom. Well-known songs of the time included "Swing Low Sweet Chariot," "Go Down Moses," and "Deep River." Blues evolved from spirituals as a more secular form of similar, and far earthier, sentiments. The blues were often about sadness, love gone wrong, and about facing insurmountable troubles. But they could also be funny, upbeat, even defiant.

Alongside the development of the blues was ragtime, a very energetic form of music powered by complicated, syncopated beats, often played on a piano. The best-known ragtime musician was Scott Joplin, who wrote many hit ragtime pieces for the piano including "Maple Leaf Rag." African rhythms, slave work songs, spirituals, blues, and ragtime eventually combined to create the ferment from which jazz would bubble up. In the late 1800s, jazz was at its beginnings, but not long after the turn of the century, it would become the most popular form of American music.

It wasn't until the parlor and minstrel songs of Stephen Foster that America began to be recognized as having its own form of music. Foster, a white man from the North, combined parlor piano music with the sadly racist blackface entertainment of the minstrel show and ended up becoming the most popular composer of the 19th century, and is sometimes referred to as the "father of American music." His songs had memorable melodies that were easy to sing and lyrics that spoke to aspects of American life of the time—songs like "Oh, Susanna," "Camp Town Races," and

"Beautiful Dreamer." In 1848, Foster sold "Oh, Susanna" for $100. It was his wife, Jane, who inspired his song "Jeanie With the Light Brown Hair." Many of Foster's songs, like "Swanee River" and "My Old Kentucky Home," idealized life in the South though he only visited it once, taking a riverboat trip down the Mississippi river.

It was frequently wartime which produced the songs that enjoyed widespread success. During the Revolutionary War, the country's favorite marching tune was "Yankee Doodle Dandy." During the Civil War, "The Battle Hymn of the Republic" was a favorite among the Union soldiers, while "Dixie" became the anthem of the Confederate troops. Another popular Civil War song was "When Johnny Comes Marching Home."

The Industrial Revolution had an impact on the design of musical instruments. Better quality and more durable materials allowed instruments to be made more cheaply and in greater numbers. Instrument designs also were improved during this time. Distinguished performers such as the violinist Niccolo Paganini (1782-1840) and the pianist and composer, Franz Liszt (1811-1886) were widely admired and treated like royalty, and their preferences changed the designs of instruments. The larger and louder violins preferred by Paganini replaced the quieter and subtler violins of earlier players. The demands of pianists like Liszt encouraged the design and development of bigger and better pianos, and eventually, to the replacement of the internal wooden structures of 18th century versions of the instrument with cast iron frames.

The Industrial Revolution also created a middle class that had more disposable income and more leisure time than ever before. There was more time for more people to learn and play music as it gained in popularity throughout the 19th century. Home life was centered on the salon or parlor room. Children learned and played there and it's where the family entertained their company. With the rise of the parlor as the center of family life, music education, which was becoming increasingly important, was centered there, too. Children were often taught to play musical instruments as part of their well-rounded education. All sorts of musical instruments began to be used in the home. The guitar, harp, concertina, and banjo were popular, but the most important musical instrument in the home was, of course, the piano. To accommodate for home use, smaller pianos were created, first square models, and later, uprights. The small pianos took up less space and although they were less powerful than their larger siblings, they were also far more affordable.

CHRISTIAN STEINWAY (1825-1889)

In 1853, Steinway & Sons was formed. Operations began in a rental loft at 85 Varick Street in New York City. The Steinway business was a family affair, with father Henry Engelhard Steinway, working as both a master cabinet maker and a master piano builder. Doretta, the eldest daughter, was the star salesperson, sometimes offering to give free piano lessons to prospective buyers in order to close a sale. Christian Friedrich Theodore Steinway was the eldest son of five boys and three girls.

When the rest of his family had originally immigrated to the United States, Theodore (as he was known) stayed behind to run his own piano factory in the family's native Germany. After the death of two of his younger brothers, Theodore was asked to give up his business and join the family business in New York as a partner. He arrived in New York in 1865 and became head of the technical department of Steinway & Sons.

Theodore was a brilliant, meticulous engineer and scientist. He leaned heavily on theoretical research in acoustics. Building on the work of his brother, Henry, Jr., he accelerated the pace of invention, research, and development within the company and was responsible for most of the major improvements associated with Steinway & Sons. The company

Figure 1: Advertisement: Steinway & Sons

licensed Theodore's patents—45 of the firm's 101 patents—for its exclusive use. The technical changes Theodore proposed in his patents gave Steinway pianos greater strength and stability, more responsive action, bigger, purer sound, greater ease of repair, improved design, and better tools for piano construction. Most changes focused on the plate, case, hammer jack and spring, and soundboard. Of the 15 Steinway patents granted before the Centennial Exhibition of 1876, 13 related to grand pianos. The 26 patents granted between 1878 and 1885 included 11 for grands, 11 for uprights, and four that either applied to both types or improved manufacturing techniques.

Theodore's success with upright pianos in Germany encouraged him to persist with their large scale production in the United States. After studying the steel industry in the United States and Europe, he succeeded in producing a cast frame which permitted string tension to be nearly doubled. Controlling string tension allows for better control of the sound quality. As an experiment, he built a parlor grand that was only six feet long but had the same power and tone as a much larger instrument. Before the end of the 19th century, the Steinway grand was substantially the instrument we know today and the upright had become the universal home piano. Theodore Steinway had created a piano which was accepted as the piano standard throughout the world.

U.S. Patent #204,111 was very significant in the history of pianos and musical instruments. "The capodastro bar was quite a revolutionary development in its time and is still very much a standard part of every Steinway grand piano. It has been copied by practically every other piano maker since the patent expired a long time ago and is now standard in every piano made all over the world," said Theodore's great-nephew, John H. Steinway.

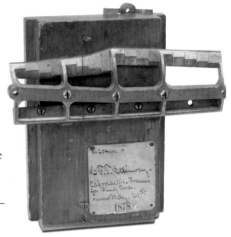

CAPODASTRO FRAMES FOR PIANO FORTES, PATENT NO. 204,111
Christian F.T. Steinway, New York, New York, May 21, 1878

The capodastro frame consisted of one up-bearing and two down-bearing strips, which were connected by cross ribs and all cast in one piece. Each piano string started at the pin block on the piano frame and traveled through the capodastro frame which was located on the treble end of the piano frame. Changes to the form and distances of the up and down bearings on the capodastro's frame ensured each string would be in harmony or in tune with the main portion of the string.

LEVI FULLER (1841-1896)

Levi Fuller, accomplished in the fields of business and politics, was also a prolific inventor. He is credited with personally receiving over 100 patents, mainly in railroad equipment and reed organ design.

Fuller was born in Westmoreland, N.H., the son of Washington and Lucinda Fuller. His parents were of English and German descent and his ancestors on both sides served in the Revolutionary War. At the age of 4, his family moved across the Connecticut River to Vermont. Always an ambitious youth, Fuller apprenticed with a machinist in Boston and became a mechanical engineer. At the age of 19, he returned to Brattleboro, VT., and became a mechanical engineer for the Estey Organ Company (Figure 2), founded by Jacob Estey. Six years later, he became superintendent of the company and also served in the roles of patent expert and inventor. In 1865, Fuller married his boss' daughter, Abby. A year after, Jacob Estey took as partners both his son, Julius, and his son-in-law, Levi. Together, they directed the company's operations until Jacob Estey died in 1890.

The sense of civic responsibility exhibited by the Estey Company was unusually progressive for its time. When the factory burned down, the company invested in a fire department. The Estey Company paid equal wages for men and women, unheard of in the late 1800s. The company needed gas piped to it, so they supplied residents along the route with their first gas lamps. Wood shavings from the carvers were recycled to help fuel the boiler in the Engine House. When a hurricane hit, the company switched its power over to help the nearby hospital.

Many of Levi Fuller's patents were assigned to the Estey Organ Company, the leading manufacturer of high-quality reed organs in the United States. Over 500,000 instruments were handcrafted in its Vermont facility and shipped throughout the U.S. and around the world.

The Estey Organ Company was at its peak in 1892. The Governor of Vermont was in attendance to commemorate the production of the company's first 250,000 organs. This event may have marked the high point of the reed organ's popularity in the U.S. as pianos soon began to displace organs, especially in larger cities.

Levi Fuller, in his roles as businessman, politician, community leader, and inventor made a significant impact in the state of Vermont. He was elected to the Vermont Senate in 1880, the lieutenant governorship in 1886, and the governorship of the state in 1892.

Figure 2: Early Advertisement Estey Organ Co.

REED ORGAN, PATENT NO. 139,666
Levi K. Fuller, Brattleboro, Vermont,
Assignor to J. Estey & Company, June 10, 1873

This set of reed organ improvements related mostly to organs with more than one bank of keys. Fuller claimed "a better combination and arrangement of parts, compactness of form and increased variety of music and power of tone." Many improvements were made, but the most notable was a lever and stop mechanism for the blowing apparatus. Throwing this lever allowed a second person near the organ to work the blower's foot pedals, freeing the performer to manipulate the foot pedal tone keys. The organ combined a lever or levers with the keys and pedal-rods to communicate the motion of the pedal-rods to the valves. The arrangement of the pedal-bass reeds was placed below the foundation board and a pivoted bar was used to hold the pins in place.

REED ORGAN, PATENT NO. 160,052
Levi K. Fuller, Brattleboro, Vermont,
Assignor to J. Estey & Co., February 23, 1875

This new reed organ allowed the musician to add a tremolo effect to any single note without adding the effect to other notes, by applying additional pressure to the desired key. The organ included a second bank of reeds that were only activated by the extra pressure. This second reed bank's sound fed into a tremolo-effect fan box that spanned the length of the reed board.

OLE BULL (1810-1880)

Ole Borneman Bull (Figure 3) was best known as a Norwegian violin virtuoso and composer, and has been called Norway's first international star. He also invented a grand piano with a new kind of sounding board.

Figure 3: Portrait of Ole Bull

The oldest of ten children, Bull was mainly self-taught. At four or five years old, he could play on the violin all of the songs he had heard his mother sing. At the age of nine, he played first violin in the Bergen Theatre Orchestra and was soloist with the Bergen Philharmonic Orchestra.

He met with great success touring the U.S. While in America, Bull became interested in founding a colony for fellow Norwegians, where they could come to seek their fortunes and make a new beginning. In 1853, he purchased 120,000 acres of land in Pennsylvania and founded a colony called New Norway. However, the project was abandoned in less than a year, as the land was not suited for agriculture. The original site chosen for Bull's colony is now the Ole Bull State Park in the Susquehannock State Forest.

Bull's talent was considered to be on a level with Niccolo Paganini, and he was a friend of Franz Liszt, with whom he played on several occasions.

PIANO, PATENT NO. 109,172
Ole Bull, New York, New York, November 15, 1870

The musician and inventor Ole Bull claimed many new improvements relating to the design of the grand piano. A major focus was a uniquely-designed soundboard and its surrounding components. He outlined the soundboard's design and specified strict adherence to guidelines for its construction, which was a matter of acoustical importance. Other noteworthy improvements included hollowed-out legs that featured valves to direct sound into the glass wheel casters, as well as a method to isolate the soundboard from the exterior piano case.

VIOLINCELLODIAN MUSICAL INSTRUMENT, PATENT NO. 20,397
John D. Akin, Spartanburg, Pennsylvania, June 1, 1858

Inside the violincellodian, a series of violin shells diminishing in size from bass to tenor were laid flat in a frame. The musician pumped the foot treadle to revolve a large flywheel whose belt drive powered two shafts. Each shaft revolved an endless leather belt coated with horse hair which acted as a bow to the strings of the violin shells. Each keyboard was assigned a violin shell, and pressing the keys lifted and lowered the shells to make contact with the revolving bow strap which played the strings.

REED ORGAN, PATENT NO. 205,341
George Blatchford, Mitchell, Ontario, Canada, June 25, 1878

This organ improved on the inventor's previous patent for a resonant chamber for reed organs. The improvements included repositioning the reed board and inserting it into grooves inside the chamber to obtain a more perfect vibration. In addition, an auxiliary front swell could be used independently or with the grand swell at the top of the chamber. These improvements resulted in a "more distinct and perfect vibration of the chamber."

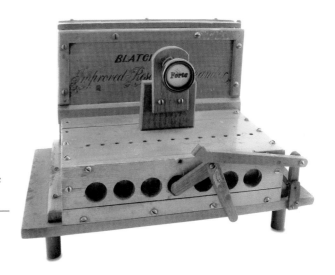

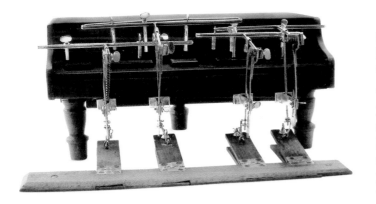

HARMONY ATTACHMENTS FOR PIANO FORTES, PATENT NO. 221,609
Gustav A. Radinsky, Marshall, Texas, November 11, 1879

The Harmony Attachment could accompany a solo musician whose hands were occupied with an instrument, such as a violin. The apparatus was placed in front of a piano and its arms adjusted to hover over the keyboard, with the racks of fingers positioned over the appropriate keys. Each pedal controlled a rack of fingers, pressing the piano keys as the corresponding foot pedal was depressed.

MUSICAL BOX, PATENT NO. 215,146
Henri J. A. Metert, Geneva, Switzerland,
Assignor to M. J. Paillard & Co., May 6, 1879

This musical box played a tune by rotating a pin-studded cylinder against a set of picks. The cylinders were interchangeable and could easily be inserted or removed. Some cylinders had more than one tune arranged on them, and this improved musical box let the user skip over tunes and select the desired one.

REED ORGAN, PATENT NO. 134,830
George Woods, Cambridgeport, Massachusetts, January 14, 1873

This organ could be played as a traditional reed organ and simultaneously as a piano-sounding instrument. The "vibratory hooks or forks" in the rear of the organ produced piano-like sound when struck by the key-actuated hammers. The piano action could be disengaged when desired.

KEYED MUSICAL INSTRUMENT, PATENT NO. 58,959
Hubert C. Baudet, Paris, France, October 16, 1863

This model is a partial representation of a keyed instrument that produced sounds similar to a bowed instrument. Each keyboard key has a roller at its tip, with a small metallic thread that would make contact with a string mounted to the rear sounding board. As a key is pressed, the key's roller moves vertically to make contact with two cylinders at the top, which rotate using power from two foot pedals. The contact between the metallic thread and the string produces a violin-like sound.

PIANO, PATENT NO. 57,186
Richard Raven, New York, New York, August 14, 1866

This piano's power and tone quality were enhanced by additional sounding boards, referred to as sounding drums or vibratory chests. The boards were placed underneath the top main bridge sounding board. It was claimed that "the vibratory power of the piano would be increased without lessening the sound from the top main sounding board."

PIANO SOUNDING BOARD, PATENT NO. 152,159
Friedrich William Niehaus, St. Louis, Missouri, June 16, 1874

This sounding board improvement applied to piano soundboards with two or more bridges. The inventor claimed that adding corresponding counter-bridges on the underside of the soundboard would produce a "strong, full, sweet, and harmonious tone."

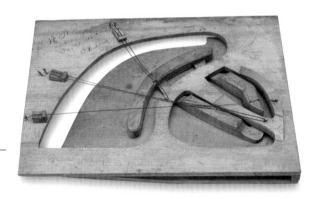

DRUM, PATENT NO. 223,586
Augustus L. Fayaux, Philadelphia, Pennsylvania, January 13, 1880

This device for tightening a drum head was said to be compact, convenient, and effective. Its threaded thumb key was hooked to the drum's rim along with an eye block where the tension cord would pass. Turning the thumb key, which remained stationary, threaded it into the eye block, drawing it upwards and tightening the tension cord.

MUSICAL ROCKING CHAIR, PATENT NO. 106,790
Clayton Denn, Philadelphia, Pennsylvania, August 30, 1870

The bellows installed under the seat of this rocking chair was pumped by rocking the chair. Reeds covered by knob-controlled valves were positioned under the arm rests of the chair, so the chair's occupant could reach and manipulate the knobs easily, playing the chair as a musical instrument. The chair also included a throwout lever to disengage the bellows when the rocking chair was not being used as an instrument.

SELF ADJUSTING OPERA SEAT, PATENT NO. 12,017
Aaron H. Allen, Boston, Massachusetts, December 5, 1854

This seat automatically folded up into a vertical position once a patron stood up from the chair. The chair hung on a shaft which acted at its center, with a spring formed around the shaft to engage and collapse the chair once the weight was removed from the seat bottom. The chair could be used singly or in rows.

KEY BOARD FOR MUSICAL INSTRUMENTS, PATENT NO. 161,806
Martin H. McChesney, Pontiac, Michigan, April 6, 1875

This keyboard's unusual construction and key layout promised a much more comfortable and efficient way to play any keyboard instrument. According to its inventor, any scale, arpeggio, chord, or melody could be played by the same manipulation of the hand. The musician would learn one musical key study and then be able to apply it to all twelve keys without having to learn twelve different positions.

DEVICE FOR TEACHING MUSIC, PATENT NO. 152,726
Francis Cramer, Toledo, Ohio, July 7, 1874

This new keyboard layout consisted of six white keys and six black keys representing one octave, instead of the more familiar seven white and five black keys, which the inventor described as "the old system." Any one of the 12 key scales could be played in the very same manner, saving time in the separate study of 12 different key position patterns. This keyboard worked in connection with a sliding scale that was meant to be repositioned to the tonic of the keyboard.

PIANO STOOL, PATENT NO. 76,044
James Bramble, Hugh M. Deihl, Fort Wayne, Indiana, March 31, 1868

This hydrostatic piano stool could be elevated or lowered using a foot pedal. Pressing with the foot pedal forced down a plunger in the liquid reservoir in the base of the stool, forcing the liquid into a chamber and pushing the seat up. By lifting the foot pedal with the heel, the liquid returned to the reservoir, thereby lowering the seat.

PIANO FORTE ACTION, PATENT NO. 12,003
Daniel H. Shirley, Boston, Massachusetts, November 28, 1854

This new Piano Forte action had improved devices for lifting the hammer and was arranged to keep the weight of the hammer always upon the key lever, so the slightest touch would "deliver a blow" regardless of the key's position. Other improvements included simplified construction to reduce cost, prevent "blocking" of the hammer at inappropriate times, and minimize adjustments. The action "combined power with lightness and a sensitive touch."

CHAPTER 22

Sports & Entertainment

"It is time for us all to stand and cheer for the doer, the achiever – the one who recognizes the challenges and does something about it."

— Vince Lombardi

In the early part of the 19th century, many leisure activities were organized around wealth, gender, class, and ethnicity. The wealthy attended theaters, restaurants, and sporting clubs. Leisure for the working class centered on ethnic communities and neighborhoods, each with their own saloons, churches, fraternal organizations, and organized sports. By the end of the 19th century, people had more free time and money to spend on leisure activities as well as greater access to inexpensive forms of transportation. A middle class was emerging, one that would become increasingly involved in sports, as both spectators and participants. During the 19th century, sports became organized. The first written rules for rugby were drawn up in 1845, the London Football Association devised the rules of football (soccer) in 1863, and in 1867, John Graham Chambers drew up a list of rules for boxing, called the Queensbury Rules after the Marquis of Queensbury. The Amateur Athletics Association was founded in 1880.

By the end of the 19th century, bicycling had become a popular sport. The involvement of both men and women contributed to its popularity, which peaked in the 1890s with the invention of the safety bicycle, so-called because it was considered safer than the high wheeler which it replaced. Shortly after the safety bicycle's invention, John Boyd Dunlop invented the air-filled, inflatable, pneumatic tire.

Parks began to be built in urban areas, at the recommendation of physicians prescribing physical exercise as a way to reduce stress. The rise of spectator sports (primarily baseball, football, and tennis) all contributed to the changing nature of leisure in the United States. The most popular spectator sport in America in the late 19th century was baseball. The National League was formed in 1876, with the first World Series Championship in 1903. Vaudeville, dance halls, and eventually, motion pictures, became popular. Such well-known performers as Will Rogers, Harry Houdini, and the Marx Brothers all got their start in vaudeville. Movies developed thanks to Thomas Edison's kinetograph, a camera that could take photographs of moving objects. Nickelodeons, the first permanent movie theaters, began to spring up as well.

VELOCIPEDE

The velocipede (Figure 1), an early form of the bicycle, derives its name from the French words meaning swift footed. This two-wheeled riding machine had stationary handles, with foot pedals attached directly to the larger front wheel. It was made mostly of wood and its wheels had iron-clad wooden rims. The combination of these materials with cobblestone roads made for a very uncomfortable ride, and the velocipede earned the nickname "bone-shaker" as a result.

The velocipede was not easy to ride; it was heavy and difficult to mount and steer, and speed was limited because one revolution of the pedals produced only one revolution of the front driving wheel.

American inventors, seeking to make cycles safer and easier to ride, proposed and patented variations of wheel arrangements, propulsion, and seating. The "bicycle craze" of 1878-1900 stimulated many innovations in mechanical design and mass production that the automobile industry later adopted.

Figure 1: Early Velocipede c. 1860

VELOCIPEDE, PATENT NO.92,808
David J. Farmer, Wheeling, West Virginia, July 20, 1869

This velocipede was designed for use on land, but could also run on water. On encountering a body of water, the rider could simply keep going without making any changes to the velocipede's mechanism. Its wheels were hollow cylinders, which allowed the velocipede to float as it was operated on the water.

ROLLER SKATE

The first patent for a roller skate was issued to M. Petitbled of Paris, France in 1819. The "Petitbled" featured three in-line wheels, a wood plate, and leather straps to attach it to the foot. Wheels, which were all the same size, were available in wood, metal, or ivory. A skater could only skate forward in a straight line; turns and curves were impossible.

On January 6, 1863, James L. Plimpton received a U.S. patent on the first modern roller skate. His skate could turn because its mechanism had a pivoting action that was dampened by a rubber cushion, which permitted the skater to turn simply by leaning weight in the direction of travel. In this manner, a roller skate could move around a floor as if on ice. This "rocking skate" design is still in use today. Plimpton developed this skate because he had taken up ice skating for exercise and wanted to continue to skate during the summer months. Plimpton, who was in the furniture business in

New York city, built a roller skating floor in his office. He opened the world's first public roller rink, which was located in Newport, R.I, and also founded the first roller skating association. The Plimpton skate is credited with revolutionizing the roller skate industry.

Plimpton's successful patent inspired many imitators and was repeatedly infringed upon. Inventors tried in many ways to duplicate the rocking action he had developed. He spent much of his time and money protecting his patent; more than 300 manufacturers were successfully sued. Of the many patents for skates in the mid-to-late 1800s, most were either unsuccessful or infringed on Plimpton's patents. Although Plimpton became wealthy from his invention, so many people infringed on his patent that his lawyers created form letters to warn offenders.

In 1869, George Stillman received a patent for his roller skate. Stillman took Plimpton's idea, discovered an improvement in one specific area, and received a patent for it. In Stillman's patent, he claimed that his innovation was the ability to adjust the strap for size. (Figure 2)

Figure 2: First U.S. patent for Roller Skate

ROLLER SKATE, PATENT NO. 90,603
George K. Stillman, Cincinnati, Ohio, May 25, 1869

This invention consisted in connecting the rollers of the skate to the foot board so that the rollers could be turned, cramped, or adjusted (in order to follow a curved track, right or left, agreeing with the bodily motion) by means of a strap or fastening device which encircled the foot of the skater, the strap being connected to the frames which contained the rollers.

When the skater bent to describe a curve, the foot rolled over the foot board and forcibly revolved the strap, which in its turn carried the levers to the opposite side, away from the body. This adjusted the rollers so that the skate followed a curved track agreeing with the inclination of the skater.

MOSES BENSINGER AND BENJAMIN F. GOODRICH

The earliest known billiard cushions were fabricated using only short, wooden, cloth-covered walls. In 1835, crude rubber from India was substituted for the cloth, but was soon replaced by "vulcanized" rubber, a more durable, temperature-resistant product created by Charles Goodyear (Chapter 12, Manufacturing).

At some point in the history of billiard cushions and rubber tires, B.F. Goodrich and Moses Bensinger combined their efforts to incorporate modern technology in the design of billiard cushions. The significance of their discoveries in the field of rubber is evidenced by the fact that vulcanized rubber is still used in the billiard cushions produced today.

CUSHION FOR BILLIARD TABLES, PATENT NO. 226,827
Moses Bensinger, Chicago, Illinois, Benjamin F. Goodrich, Akron, Ohio, Assignor to J.M. Brunswick And Balke Company, April 27, 1880

In the past, billiard cushions were made from uncovered vulcanized rubber which allowed the ball to sink too far into the cushion. This invention overcame the difficulty by encasing the cushion in canvas or another closely woven fabric. This covering made the action of the cushion quicker, allowed the ball to rebound farther, and lessened the chance of the ball jumping off the table.

BENJAMIN FRANKLIN GOODRICH (1841-1888)

B.F. Goodrich (Figure 3) was born in the small farming town of Ripley, NY, orphaned at eight years old, and raised by his uncle. During the Civil War, he served as a battlefield surgeon for the Union Army.

Goodrich was the first man in Akron, OH to own a telephone, a gift to him from Alexander Graham Bell in 1877. The phone connected Goodrich's house on Quaker Street with his factory on Rubber Street.

Goodrich was both a businessman and a physician. In 1869, he invested in the Hudson River Rubber Company and would go on to become its president. The company's products included billiard

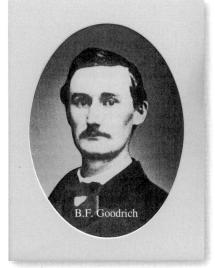

Figure 3: Portrait of B.F. Goodrich

cushions, bottle stoppers, rubber rings used for canning jars, and fire hoses. The Hudson River Rubber Company eventually became the B.F. Goodrich Company and was incorporated in the state of Ohio.

Goodrich produced the first automobile tires in the U.S., in 1896, the first rubber sponge in 1902, and aircraft tires in 1909. All American aircraft used in World War I sported B.F. Goodrich tires, and Charles Lindbergh's *The Spirit of St. Loui*s also took flight on tires manufactured by B.F. Goodrich. During World War II, Japan controlled the supply of natural rubber, so B.F. Goodrich Company invented synthetic rubber to supply the U.S.'s war needs.

MOSES BENSINGER (1839-1904)

For 150 years, the Brunswick Company was the most prominent billiard company in America. From the 1800s to the end of World War II, it remained virtually unchallenged in its leadership of the American billiard industry. John Brunswick started the business as the Cincinnati Carriage Making Company in 1845, which later would become known as J.M. Brunswick Company. In 1872, the company's name was again changed to the J.M. Brunswick Billiard Manufacturing Company. In 1890 Brunswick turned over the presidency of his company to his son-in-law, Moses Bensinger (Figure 4). Bensinger helped expand the Brunswick Company by joining forces with their competitor, Julius Balke's Great Western Billiard Manufactory. Under

Figure 4: Portrait of Moses Bensinger

Bensinger's leadership, the company experienced a period of growth and success. An accomplished billiard player himself, Bensinger spent time experimenting and researching better ways to make tables and equipment. His contributions were also significant in the bowling industry. Bensinger oversaw the company's initial foray into the bowling business. He helped organize the American Bowling Congress (ABC) and the equipment used during the first national championship. Brunswick equipment was used in all ABC tournaments for the next 40 years. At least five important patents for rubber cushions were registered in Bensinger's name, all assigned to the Brunswick Company.

AUTOMATIC TOY GYMNAST, PATENT NO. 182,194
William L. Hubbell, New York, New York, September 12, 1876

The gymnast's hand-over-hand movement along the rope was accomplished by a clockwork mechanism combined with actuating cranks. An escapement wheel and pallets within the clockwork mechanism regulated the gymnast's speed. The gymnast could operate both below or above the rope.

UPRIGHT STANCE

PIGEON STARTER, PATENT NO. 159,846
Henry A. Rosenthal, Brooklyn, New York, February 16, 1875

At the time this patent was issued, live pigeons were used for target practice, and placed in traps dug into the ground, just below grade level. However, opening the trap was often not enough to make the pigeons fly up or even leave the trap. Yelling and throwing stones at the pigeons were common methods of starting the pigeons, but affected the shooter's concentration. This pigeon starter made a loud noise and included a cat-like figure that moved from a crouching position into an upright stance, to startle the birds into flight.

Soon after this patent was issued, the shooting of live pigeons for sport was banned. The live pigeons were replaced by a patented clay disc that were then named "clay pigeons." The sport took on the name of "trapshooting" resulting from the fact that the pigeons were originally kept in the ground in traps.

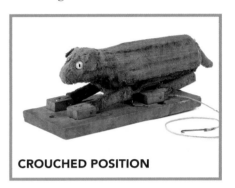

CROUCHED POSITION

EXERCISING APPARATUS, PATENT NO. 63,846
Benjamin F. Brady, New York, New York, April 16, 1867

This apparatus was "designed to furnish a means of physical or gymnastic exercise of a character similar to that of rowing." The construction of this apparatus provided a convenient way to exercise indoors with a beneficial effect to that derived from rowing on a body of water, which was a popular activity at the time.

ICE SKATE, PATENT NO. 27,193

Enos B. Phillips, Cambridgeport, Massachusetts, February 14, 1860

This invention aimed to produce a cheap, durable and handsome skate blade. Previous ice skates had been cast of iron, which became too brittle in the frost. Their runners wore out quickly, becoming dull and bruised. Phillips' skate used an alloy of copper, tin, zinc, and antimony to create a hard, tough, and therefore very durable skate blade.

MACHINE FOR CUTTING ENDS OF BILLIARD CUES TRUE, PATENT NO. 24,625

Ira Glynn, Mikel Borowsky, Placerville, California, July 5, 1859

To keep the leather cue tip on a billiard cue, the end of the cue had to be cut true. This machine for cutting the cue end consisted of a bed with a lever for attaching a cutter or knife. The groove in the block and bed piece received the cue, and the jaws opened to hold the cue in place so its end could be cut off square.

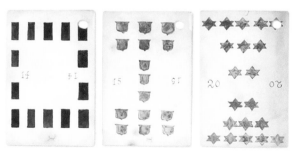

PLAYING CARDS, PATENT NO. 178,863

Isaac Levy, New York, New York, June 30, 1876

These playing cards increased and varied the combinations, especially numerically, that could be played by a regular pack of playing cards. The face cards (jack, queen, and king) were replaced by an eleven, twelve, and thirteen card. These new cards allowed a variety of new games to be played, to "vary and increase the interest in those games being played during this period of time."

GAME INDICATOR, PATENT NO. 213,113
Abner H. Jones, Ilion, New York, Isaac Osgood, Utica, New York, March 11, 1879

This invention made it easier to keep track of the order of play in games such as croquet. It consisted of a rod and a set of balls of different colors, which could be arranged in any order desired. The balls could also be marked with numbers instead of colors.

CONSTRUCTION OF CHECKERS, PATENT NO. 72,670
J. V. Henry Nott, New York, New York, December 24, 1867

This checker was wider at the top than the bottom, so it would run around in a small circle if dropped on the floor, instead of rolling away and getting lost. It also featured a rim or flange on top, to keep it from slipping off the checker placed on top of it when a king was crowned.

TREATMENT OF WOOD FOR THE MFG. OF DOMINOS, PATENT NO.133,697
George H. Chinnock, Brooklyn, New York, December 10, 1872

Previously seasoned wood is cut crosswise of the grain to produce dominos both for use and ornament. An adhesive compound was used to glue a veneer to the wood domino, after which a die or punch pressed a design into the face of the wood block. The impressed material being below the face of the block allowed the domino to be polished resulting in a smooth face.

BALL GAME, PATENT NO. 123,442
Edward A. Barrett, New York, New York, February 6, 1872

Barrett, a Commander in the United States Navy, invented a unique ball game. The playing table had pockets for the balls, as in billiard tables, but the pockets were masked by cushions. Holes in the center of the table were connected to the pockets by concealed conduits. Many complex mathematically-based games could be played, depending upon the values assigned to the five pockets on the table.

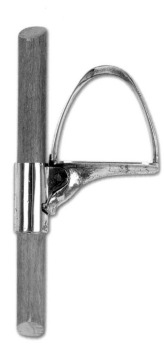

STILTS, PATENT NO. 168,213
Franklin Beaumont, Jr., Dallas, Texas, September 28, 1875

These adjustable stilts could be lengthened or shortened by moving the stirrup up or down on the stilt rod. The mechanism was a spring-loaded serrated arc-shaped blade that projected into the stilt bar. The stilt walker's weight put pressure on the serrated blade to hold it firmly in place on the stilt rod allowing the height of the stilt to be adjusted to any length desired.

GAME, PATENT NO. 82,370
William H. Wilson, Providence, Rhode Island, September 22, 1868

This parlor or field game relied partly on skill and partly on chance. The game used a revolving pointer moving over a disk on which numbers or words were marked. As the ball struck the pointer, it swung around and came to rest on a figure or number, thereby indicating the progress of the player in the game. The player who first succeeded in passing around the circle of the dials and then hitting the center post was declared the winner.

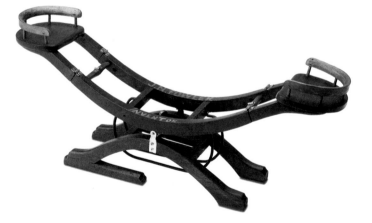

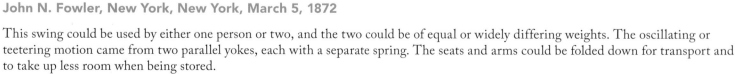

OSCILLATING SWINGS, PATENT NO. 124,262
John N. Fowler, New York, New York, March 5, 1872

This swing could be used by either one person or two, and the two could be of equal or widely differing weights. The oscillating or teetering motion came from two parallel yokes, each with a separate spring. The seats and arms could be folded down for transport and to take up less room when being stored.

MANUFACTURE OF CANDY WHISTLES, PATENT NO. 119,392
Augustus Neuhausen, Wheeling, West Virginia, September 26, 1871

In the past, candy whistles had been cast in two pieces and then connected. These improvements to the mold design and to the process meant that the whistles could be cast in one piece, saving both time and labor and thus reducing cost.

TOY PUZZLE, PATENT NO. 47,552
H.C. Ketcham, Bloomfield, New Jersey, May 2, 1865

Its inventor called this toy puzzle "The Mystic Cord." The cord was concealed in the sides of two blocks and connected at the bottom of the blocks. To demonstrate the illusion, the cord was pulled back and forth through the blocks, to prove that the cord was undivided. Then a knife blade was drawn between the blocks, to make it look as if the cord had been cut. The blocks were then rotated in opposite directions, to show that the cord had been cut. Finally, the blocks were brought back together and the cord was pulled back and forth, just the same as before it had been "cut."

CHAPTER 23

Transportation

"The American people have done much for the locomotive, and the locomotive has done much for them."

— James A. Garfield

In America during the 18th century, rivers had been the most widely used method of transporting goods and people. The roads at the time were largely based on existing Native American trails and trade routes.

Improvements in transportation were a major reason for the growth of the U.S. in the 1800s. Many more roads were built and construction of bridges increased. On the water steamboats used the river system, its superhighway being the Mississippi River.

The covered wagon was in full use at this time and would become the iconic symbol of westward migration and pioneer settlement. The all-purpose land ship for handling overland freight was the Conestoga wagon, first built in Pennsylvania's Conestoga Valley. Before the canals and railroads, Conestogas in caravans of 10 to 12 carried most of the heavy freight to the West.

Gradually, the country began to develop an organized system for transporting people and goods. The idea of using federal funds to build roads was established in 1806 under Thomas Jefferson. The National Road, which stretched from the Atlantic seaboard to the Ohio Valley, was begun in 1811. This road increased business for the city of Baltimore while taking business away from New York City. To compete, New York built the Erie Canal, connecting the Great Lakes with the Atlantic and establishing the Port of New York as a major trading center. The success of the Erie Canal benefited the cities along its path. Canal building peaked around 1836. Canal traffic began to dwindle in the 1860s as railroads gained in popularity. (See Chapter 15, Marine & Navigation.)

Railroads were initially seen as a way to compete with the Erie Canal to transport goods. They grew rapidly throughout the mid-19th century, tying together much of the country, particularly the Northeast and Midwest. The growth of railroads had a large influence on the growth of industry. Because large amounts of iron and steel were needed for railroads, the production of iron, steel, and coal all grew. The rise of cheaper land transportation allowed farmers to sell their crops to larger markets over greater distances. Cheaper travel permitted more people to travel while travel time decreased. The growth of cities and towns was determined by their nearness to railroads.

In 1862, Congress passed the Pacific Railway Act. This Act authorized the construction of a transcontinental railroad. It designated two companies to build and operate a railroad between the Missouri River and Sacramento, the Central Pacific from the West, and the Union Pacific from the East. The first railroad under the Act was completed on May 10, 1869. By 1900, four additional transcontinental railroads connected the eastern states with the Pacific Coast. The transcontinental railroads linked the two coasts and opened the West to safer, easier, and more rapid settlement. The railroads moved coal for steam power, heat, and gas lighting for use in the cities, petroleum from Pennsylvania, timber from Washington State, perishable fruit from California, potatoes from Idaho, and wheat from Kansas. Suddenly, many varieties of products were available to people and businesses throughout the country.

In the early years of the railroad boom, British locomotives were imported into the U.S., but American manufacturers soon began to build locomotives in this country. Peter Cooper built the first U.S. locomotive in 1830. His steam locomotive, known as "Tom Thumb," ran on the Baltimore and Ohio Railroad. The swivel truck for locomotives was devised by John B. Jervis in 1831.

Inventions like automatic couplers, telegraphic dispatch, and air brakes contributed to the increased safety of rail travel. Rail car comfort improved with each decade. In the beginning, cars were heated by a stove in each coach. After 1890, steam heat from the locomotive circulated throughout the train. In 1864, George Pullman built a new kind of train car. The Pullman car featured plush upholstery, clean sheets, and washrooms. Pullman introduced the idea of luxury to travel.

George Pullman's Palace Car Company was incorporated in 1867. Pullman applied for patents on a hotel car with sleeping berths, dining rooms and staterooms. Pullman won the sleeping car contract for the Central Pacific Railroad.

Railroads were usually planned by private investors often with local or state government aid. Some states, including Pennsylvania, Virginia, and Georgia built and owned their own railroads.

Because American roadways and roadbeds were built as cheaply as possible and had steeper grades and sharper curves than English railroads, they needed more powerful locomotives. The American type locomotive, the 4-4-0, was the mainstay of American railroads during much of the 19th century. At the front of the locomotive there was usually a cow catcher, a device attached to the front of a train to clear obstacles like animals or debris off the track. Next came a swivel or bogie truck, whose four leading wheels swiveled independently and followed the track as it curved, preventing the locomotive from derailing, and then four rigidly coupled drive-wheels with no rear truck. The engine and fireman stayed in a cab at the rear of the locomotive, just behind the boiler. The locomotives burned wood, not coal, because of the easy availability of wood. Wood was stored in a tender car pulled directly behind the locomotive. The locomotives required spark-arresting smoke stacks, to avoid starting fires along the route. Many patents were granted for different stack designs.

By 1876, railroads were the dominant type of long distance land transport in America. They provided many new economic opportunities and led to the growth of many towns and communities along their routes. The great success of the railroads and their acceptance by the public was due to the fact that they were safe, economical, reliable, and fast.

SAMUEL GEOGHEGAN (1844-1928)

Born in Dublin, Ireland, Samuel Geoghegan was a draftsman and mechanical engineer employed by the Guinness Brewery there. He developed several small steam locomotive engines to work the narrow gauge track he had designed and built in the brewery. He also designed special unique convertor wagons, which allowed the narrow gauge locomotives to travel on the broad gauge railway adjacent to the brewery premises. (Figure 1)

Figure 1: Guinness Tramway c. 1920

DRIVING MECHANISM FOR STREET LOCOMOTIVES, &c., PATENT NO. 226,230
Samuel Geoghegan, Dublin, Ireland, April 6, 1880

This invention relates to improvements applied to locomotive engines, tram cars, and other vehicles propelled by gas, steam, or other motive power, whereby the driving mechanism is removed from proximity to dirt and dust. The cylinders and crankshaft were positioned on top of the boiler in a horizontal position. This position imparted motion to the axle of the driving wheels by a crank and rod directly connected from the crankshaft above.

HAKON BRUNIUS (1842-1902)

Swedish inventor Hakon Brunius (Figure 2) was born in 1842. As a telegraph engineer in Jonkoping, Brunius ran a small experimental workshop where he developed what he called a railway protector. He also conducted experiments with electricity, constructing Sweden's first generator and manufacturing the first Swedish light bulbs. He was referred to as "Sweden's first electrician." He developed an early piece of safety equipment for railway trains and fire alarms. Brunius made magnet telephones with lathed wooden covers and

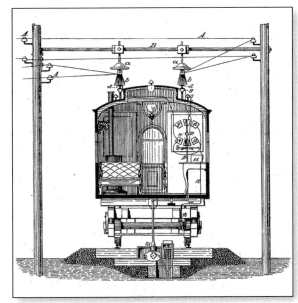

Figure 2: Brunius' Patent Drawing—Signaling Device for Railroad Car

developed wall telephones with both inductor and ringing devices, along with a wall telephone with a handheld micro-telephone.

Fixed telephone lines were installed in many parts of Sweden in 1878 and 1879. On January 3, 1878, Brunius completed the installation of a fixed line telephone connection between the head office of Munksjo paper mill and its location in the city. Ahead of his time, he proposed the establishment of a telephone network in Stockholm that was considered too novel and was never adopted.

ELECTRO MAGNETIC RAILROAD SIGNAL, PATENT NO. 189,999
Hakon Brunius, Jonkoping, Sweden, April 24, 1877

This improvement was a means to properly control and monitor the movements of trains by registering their departure and arrival times. It consisted of three parts, the station apparatus, train apparatus located on the train, and road apparatus all working together. The invention relied on two ordinary telegraph lines, a train being equipped with a means to physically contact the telegraph line as it passed, and a contact point connected near the rail of each station. This model represents the main train station apparatus to be used.

T. B. WHITE (1807-1885)

In the late 19th century, hundreds of bridge building companies were designing, building, and selling metal truss as well as other types of bridges. Bridge companies established business partnerships with foundries and metal manufacturers for cheap supplies of iron and steel, the raw materials of metal truss bridges. The companies formed, drilled, and assembled the pieces in their own shops, i.e. "prefabricating," prior to shipping them to the bridge site, where they were erected onsite.

Figure 3: Bridge Across Monongahela River at Glenwood, PA. Erected by Penn Bridge Co.

Pennsylvania, with its plentiful steel and iron mills and vast network of railroad lines, was a prime location for a bridge-building industry. Between 1870 and 1900, more than 100 bridge fabrication companies had operations there.

The Penn Bridge Company of Beaver Falls, PA was organized in 1868 as T.B. White & Sons. At the time, the firm constructed wooden bridges in New Brighton, PA. It moved across the Beaver River to Beaver Falls in 1878. In 1887, the firm reorganized and incorporated as the Penn Bridge Company, producing wrought iron, steel, and combination bridges. The company also constructed iron substructures, buildings, roof trusses, and architectural ironwork, but was best known as a fabricator of metal truss bridges. All truss bridges work by transmitting the roadway load to the abutments, but can differ from one another in the way their members are arranged. The various patterns of the 19th century were named after the engineers who devised them (Pratt, Warren, and Whipple trusses) or after the shapes formed by the upper and lower chords (bowstring, arch, lattice, lenticular, K, to name a few). Later, the invention of the automobile meant that highway trusses had to be designed for increasingly heavy loads.

Many examples of truss bridges made by the Penn Bridge Company have been preserved and can still be seen today. (Figure 3)

METALLIC SHOE FOR THE BRACES OF TRUSS GIRDERS, PATENT NO. 20,011
T. B. White, New Brighton, Pennsylvania, April 20, 1858

This invention focused on the bracing points found at the base of bridge truss girders. The cast iron brace, or shoe, came in two pieces similar to a tongue and groove so they could interlock at the lower string pieces and the posts. Once the two pieces were joined, a space allowed wedges to be driven into each side of the shoe, giving the girder some camber to raise it if it had been set too low.

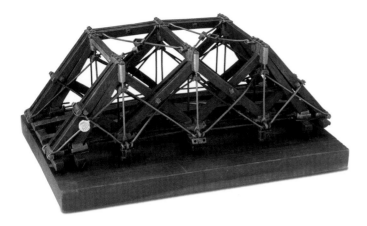

TRUSS BRIDGE, PATENT NO. 205,799
William Ireland, Oak Springs, Iowa, July 9, 1878

This invention improved on the design of truss bridges, by doubling the brace work sections throughout and forming two separate parts in each section. Its strength could be increased by the addition of more connecting rods. The bridge was designed so any brace or bolt could be replaced or repaired without needing to set up temporary trestle work to support the bridge.

DEVICE FOR OILING STEAM VALVES, PATENT NO. 225,912
George W. Baker, Erie, Pennsylvania, March 30, 1880

This oiling device automatically dispensed oil to steam engine valves when there was suction produced by the steam chest. Without suction, the device's valves automatically closed, preventing oversaturation and waste. This oiling device could be located near the engine of the locomotive steam chest or, more conveniently, inside the engineer's cab where it could be easily regulated.

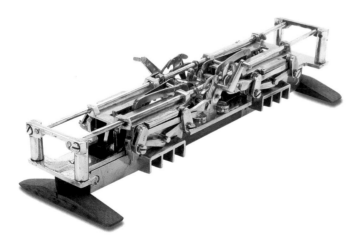

PROPELLING APPARATUS, PATENT NO. 89,435
Peter Robert, New York, New York, April 27, 1869

This invention updated the machinery for the propelling apparatus of canal boats. The apparatus had a series of floats which elevated and lowered to move the canal boat forward, like a crank paddle. The floats were raised and lowered only at the end of each stroke, and moved in a horizontal direction during the strokes.

AUTOMATIC CAR BRAKE, PATENT NO. 223,317
William L. Card, Jay Noble, St. Louis, Missouri,
Assignors to Card Automatic Brake Company, January 6, 1880

This automatic train brake patent included many improvements, most notably, its automatic effect, which was produced by a weighted revolving bar or frame that pivoted to a sleeve which both surrounded and clamped to the axle. The axle's endwise sliding collar was actuated by the revolving bar and effected by centrifugal force. The brake also featured a direct connection between the car's draw bar and the braking system, engaging and disengaging the brakes depending on its position.

CAR STEP, PATENT NO. 227,310
Benjamin F. Shelabarger, Hannibal, Missouri, May 4, 1880

This railroad car step could flip up or extend downward to allow passengers easier access when boarding or exiting a train car. When flipped up, the treads of the steps were protected, helping to prevent any potential damage caused by sparks, cinders, or snow while the train was in motion.

FARE REGISTER AND RECORDER, PATENT NO. 326,720
Joseph Corbett, Buffalo, New York, September 22, 1815

This register for tallying fares was specifically designed to prevent fraud on the part of conductors or collectors. The register included three ways to tally. The first was a daily counter that was clearly visible through a window and sounded an alarm bell after each count. Second, the machine dispensed a paper punched daily tally. Finally, the machine included a permanent register that counted all the collected fares over its lifetime, and could never be reset.

MOUNTED HORSE POWERS, PATENT NO. 202,862
Marcellus H. Pitts, Chicago, Illinois,
Assignor to H.A. Pitts' Sons Manufacturing Company, April 23, 1878

This invention improved the pinion gear frames so that different sized pinions could easily be changed out to address the machine's gearing needs. A new braking system was activated by a lever that pressed down a special "brake-band" that passed halfway around the pinion to apply a degree of friction.

LOCOMOTIVES, PATENT NO. 147,927
David Frazer, Newburgh, New York, February 24, 1874

This new way to construct and operate a locomotive for railroads introduced novel ways to propel, turn, and stop. The locomotive's "turntable" was engineered to round a curve without swinging the axle, so as to avoid the wheels slipping on the track. Clutch couplings activated by shifting levers could move the vehicle forward, in reverse, or stop it by engaging both simultaneously.

ROADWAY AND TRAMWAY, PATENT NO. 223,431
Cornelius Bremerkamp, 107 New Cross Road,
London, England, January 13, 1880

This invention improved the durability of wood block roads. Once the road was surveyed, the blocks were to be factory pre-assembled into individual square yard sizes and later laid out on the roadway. Sheet iron was slipped in between the blocks and all was held together by rods and bolts. For tramways, railway track could be secured to the roadway.

VEHICLE SPRING, PATENT NO. 225,925
George N. French, Grafton, Vermont, March 30, 1880

This new vehicle carriage spring claimed to have an improved elastic support and a simpler design than many of the suspension types used at the time. It used side springs connected to a perch frame which then connected to bow springs.

ENGINEER'S VALVE, PATENT NO. 475,695
Albert P. Massey, Watertown, New York,
Assignor to The New York Air Brake Company, May 24, 1892

This valve improved on the air brakes found on train cars by better applying and controlling the flow of air from the pressure source. The valve handle design let the Engineer choose positions to control the brake actions, such as releasing of the brakes, stopping temporarily during service stops, or applying the brakes in an emergency.

INDICATORS FOR TRAINS, PATENT NO. 216,862
Henry C. Keyes, Philadelphia, Pennsylvania, James L. Smith,
Wilmington, Delaware, June 24, 1879

This improved message board for train stations informed passengers of train destinations and departure and arrival times. The device consisted of a series of rotating slats marked with names of stations and the expected arrival and departure times. A special wand or stick was used to turn the slats individually as needed, and to change the clock hands for the arrival or departure times.

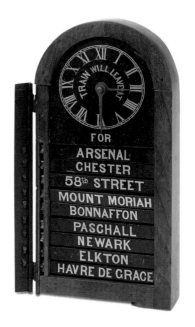

PAVEMENT FOR STREETS, PATENT NO. 222,025
Emile Devilliers, Paris, France, November 25, 1879

This paving method was intended to alleviate the deterioration of stone pavements. Paving blocks were laid down between inverted "T" bars which were fastened to transversing cross ties embedded in sand. Each block was held in place by the "T" bars and finished off with a metal strip inserted between the blocks to prevent the lateral displacement common on heavily-travelled stone roads.

CAR TRUCKS, PATENT NO. 202,737
Enoch J. Marsters, Stockton, California, Robert A. Fisher, Sacramento, California, April 23, 1878

This improvement in the wheels of car trucks let the wheels conform to a curve in a track and then return on a straightway. On a six-wheeled car, the front and back wheel and axles had wheel blocks positioned near each wheel. The wheel blocks had a obliquely-made rib tongue and groove to shift and slightly turn as the car moved along a curve. The center wheel block had a straight rib and groove to allow sideways movement.

CAR COUPLING, PATENT NO. 369,195
Madison J. Lorraine, Charles T. Aubin, St. Louis, Missouri, August 30, 1887

This improved car coupling used a one-piece automatic "gravity pin" to both release and lock the coupling "head." When coupling one car to another, the mating couplers engaged, allowing the "gravity pin" to drop down and automatically interlock the cars. To disengage, a manual lever was positioned forward, lifting the pin up and out allowing the couple head to swing open automatically due to its own weight.

ELEVATED RAILWAYS AND CARS, PATENT NO. 222,647
Edward Andrews, Pottsville, Pennsylvania, December 16, 1879

This invention attempted to lessen the noise produced by the elevated railways of the time by using a specialized railway and car. The railway track would be similar to an "H" beam in profile. Instead of wheels, the car would have a set of self lubricating "shoes" that could slide along the track and turn to conform to any track turn radius. The car also had simple gripping devices to attach to an endless cable towing system.

What's That Model ?

We hope you enjoyed learning more about the colorful, fascinating, and inspiring history of U.S. patents and patent models. We thought it would be fun to end on a little visual quiz. Following are images of some of the inventions you've learned about in this book. Can you identify them? Take this quiz and see how you do. (Answers follow, but please don't peek until you are done!)

1

A. Sextant
B. Surveying Instrument
C. Cross Bow
D. Fabric Measuring Device

2

A. Twine Maker
B. Key Duplicator
C. Candy Spinner
D. Apple Corer

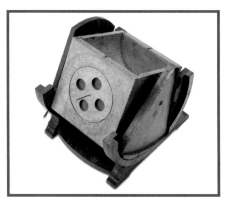

3

A. Circumference Measuring Device
B. Heliometer
C. Rocking Cradle
D. Coin Sorter

4

A. Peat Machine
B. Bingo Cage
C. Spice Mixer
D. Ore Separator

5

A. Wire Stretcher
B. Yarn Stretcher
C. Stump Puller
D. Exercising Machine

6

A. Cookie Cutter
B. Screen Splicer
C. Pasta Press
D. Perforating Stamps

7

A. Orange Juice Squeezer
B. Tea Pot
C. Measuring Funnel For Liquids
D. Milk Separator

8
A. Leather Softener
B. Boot & Shoe Cleaner
C. Callous Remover
D. Horse Brush

9
A. Paper Clip Machine
B. Circular Saw
C. Metal Grinder
D. Rotary Engine

10
A. Steam Engine Governor
B. Pop Corn Maker
C. Pressure Cooker
D. Meat Grinder

11
A. Boot Stretcher
B. Pipe Bender
C. Belt & Buckle Maker
D. Horse Collar Block

12
A. Telegraph
B. Wire Calibrator
C. Steam Engine Regulator
D. Thread Making Machine

13
A. Paper Punch
B. Needle Maker
C. Sewing Machine
D. Engraving Machine

14
A. Water Softener
B. Pipe Threader
C. Steam Trap
D. Locomotive Steering Controller

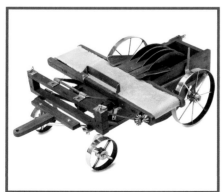

15
A. Produce Sorter
B. Ditching Machine
C. Canvas Maker
D. Battle Field Ambulance

16
A. Pasteurizing Machine
B. Alcohol Still
C. Pressure Cooker
D. Carburetor

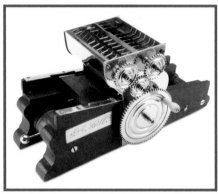

17
A. Tea Bag Maker
B. Brick Machine
C. Coin Sorter
D. Bullet Making Machine

18
A. Thread Making Machine
B. Wire Spooler
C. Telegraph
D. Cigar Roller

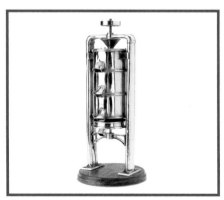

19
A. Coffee Grinder
B. Expresso Machine
C. Cider Press
D. Machine For Cleaning Semolina

20
A. Burglar Alarm
B. Flour Sifter
C. Composting Machine
D. Ice Cream Maker

21
A. Peanut Roaster
B. Gold Smelter
C. Gas Retort
D. Pipe Cutter

22
A. Cotton Gin
B. Washing Machine
C. Adjustable Desk
D. Dresser Bedstead

23
A. Paper Hole Punch
C. Safe
C. Stamp Cancelling Machine
D. Ballot Box

24
A. Furnace Flame Controller
B. Machine For Preparing Pottery Stock
C. Fire Extinguisher
D. Steam Injector

25
A. Rain Gauge
B. Oil Reservoir
C. Spittoon
D. Smoke Stack

How did you do?

IF THE NUMBER OF QUESTIONS YOU ANSWERED CORRECTLY IS BETWEEN:

0-5 > *The authors would be pretty steamed if they knew you did this bad!*

6-10 > *It's patently obvious you need to read the book again.*

11-15 > *Your gears are turning now!*

16-20 > *You can't fabricate test scores like this!*

21-25 > **CONGRATULATIONS!** *What are YOU going to invent?*

?	Answers	Invention	Patent #	Inventor	City	State	Date
1	B.	Surveying Instrument	219,076	William Devault	Cope	OH	09.02.1879
2	C.	Candy Spinner	165,836	Stewart B. Hymer	Terre Haute	IN	07.20.1875
3	B.	Heliometer	77,324	Conrad Friedrich L. Risch	Huntingburg	IN	04.28.1868
4	A.	Peat Machine	146,746	Leander W. Boynton	Hartford	CT	01.27.1874
5	C.	Stump Puller	235,842	William Youngblood	Cedar Springs	MI	12.21.1880
6	D.	Perforating Stamps	209,915	Henry H. Norrington	West Bay City	MI	11.12.1878
7	C.	Measuring Funnel for Liquids	90,457	G.B. Massey	New York	NY	05.25.1869
8	B.	Boot & Shoe Cleaner	210,072	G. Frederick Ziegler	Jersey City	NJ	11.19.1878
9	D.	Rotary Engine	214,702	Tapley B. Pyron	Springfield	MO	04.22.1879
10	A.	Steam Engine Governor	92,051	William H. Howland	SanFrancisco	CA	06.29.1869
11	D.	Horse Collar Block	13,132	Peter Moodey	Indianapolis	IN	06.26.1855
12	C.	Steam Engine Regulator	204,828	Caldwell C. Jenkins	Philadelphia	PA	06.11.1878
13	C.	Sewing Machine	19,723	James & Amos Sangster	Buffalo	NY	03.23.1858
14	C.	Steam Trap	143,761	James W. Hodges	New York	NY	10.21.1873
15	B.	Ditching Machine	198,785	Matthew J. Austin	Bonham	TX	01.01.1878
16	D.	Carburetor	151,625	John Ruthven	Point Levi	Canada	06.02.1874
17	B.	Brick Machine	15,808	Jos. A. Hill	Greencastle	IN	09.30.1856
18	C.	Telegraph	114,771	Ludovic Charles Adrien Joseph Guyot D'Arlincourt	Paris	France	05.16.1871
19	D.	Machine for Cleaning Semolina	213,284	Carl Haggenmacher	Pesth	Austria	03.18.1879
20	A.	Burglar Alarm	92,610	Andrew George Hutchinson	Stoneycroft near Liverpool	Great Britain	07.13.186
21	C.	Gas Retort	145,267	James Wotherspoon & William Foulis	Glasgow	North Britain	12.02.1873
22	B.	Washing Machine	193,732	Thore E. Smitback	Utica	WI	07.31.1877
23	C.	Ballot Box	208,951	Willis L. Barnes	Charlestown	IN	10.15.1878
24	B.	Machine for Preparing Pottery Stock	138,824	Samuel R. Thompson	Portsmouth	NH	05.13.1873
25	D.	Smoke Stack	226,439	Huntington Brown	Mansfield	OH	04.13.1880

Six Patent Models You Can Make

"There are no rules. That is how art is born, how breakthroughs happen. Go against the rules or ignore the rules. That is what invention is about."

— Helen Frankenthaler

Electromagnetic Motor

The inspiration behind this, replica comes from the "Electro Magnetic Engine" (Chapter 18, Miscellaneous, patent# 122,944), an 1872 patent by Charles Gaume. The original model ingeniously rotated iron bars by sequencing electromagnetism through the coils, converting electrical energy into mechanical energy as the magnets shut on and off to keep the wheel rotating. This simple mechanism is still the foundation for how most electric motors operate today.

HOW TO REPRODUCE THE ELECTROMAGNETIC MOTOR

REPRODUCTION BY MICHAEL CURRY

Bill of Materials

SUPPLIES AND SPECS

- Printed parts located here: http://www.thingiverse.com/thing:623901
- 4x #112 U-bolts, complete with hardware
- 2x 608 Bearings (also called skateboard bearings)
- 3x M5x10 or similar bolts.
- A spool of 22 gauge wire (solid core preferred, but stranded wire will work)

TOOLS

- Hacksaw
- Vise

Figure 1: 3D Printed parts

Section One – The Electromagnets

The three electromagnets are the heart of this electromagnetic motor model. Each electromagnet is made from a U-bolt, 22 Gauge core wire, and 3d printed parts.

The U-bolt is a #112, or U-bolt for 3/8th inch pipe. They can be found at most home centers and hardware stores, and come as a complete set with the U-bolt, two nuts, and metal plates. *You'll need all of these parts to build the motor, so make sure you buy enough for four sets.*

1. The U-bolts need to be shortened before we can use them in the motor. In the set of printed parts there is a simple cutting jig that simplifies the task.

2. Press the cutting jig down into the U-bolt so it is firmly seated.

3. We need to cut off the ends of U-bolt so they are flush with the end of the cutting jig. The best way to do this is to clamp the bolt and jig into a bench vice and cut the ends flush with a hacksaw.

4. You need to trim both legs of the U-bolt.

5. Repeat this process and trim two more U-bolts.

6. You can clean up the cut ends of the U-bolts with a metal file or bench grinder. Cleaning up the ends will make it much easier to screw the nuts on later.

7. There are two halves to each electromagnet support and a bolt that holds them together. I used M5 x15's, but any bolt that holds everything together will work.

8. Clamp the trimmed U-bolt between the two parts of the support and bolt them tightly together.

9. Now we are going to start winding the coil of the electromagnet. Begin by screwing the nuts onto the U-bolt so they are flush with the cut ends. As you work make sure to leave long leads so you don't have to splice wires later.

10. When winding a U-shaped electromagnet, the orientation of the wires is important. Start the first side of the electromagnet by passing the wire behind the U-bolt and wrap it around clockwise. Wind the wire tightly and neatly, we're going to be putting a lot of it on.

11. Keep winding the wire until you get to the nut, then start winding a second layer, moving back down the bolt. Always keep wrapping in the same direction.

12. Keep winding the wire and creating new layers until you have a coil that is four layers deep. This is three layers.

13. The fourth and final layer.

14. After the fourth layer, bring the wire across to wind the other side of the electromagnet. This side of the electromagnet needs to be wound in the reverse direction. Note the way the wire crosses the base before wrapping.

15. This side of the electromagnet needs to be the four layers deep. But unlike its brother, it is wound in the opposite direction.

16. Once you have both coils wound, test the electromagnet by powering it from a 1.5 volt battery and see if it picks up something.

17. Repeat these steps to complete the other two electromagnets.

18. Wrap the coils in tape or heat shrink to keep them from unwinding.

Section Two – The Rotor

The electromagnets are powered to attract the ferrous brackets in the rotor. *Note: Optionally, you can sand blast the brackets to remove the shine, giving them a more antiquated look.)*

1. Press each of the brackets into the rotor wheel slots. You may need a knife to shave off some plastic in the slots to fit the brackets snugly. If the brackets don't stay in place on their own, use a little glue to secure them.

2. Install all four brackets in the rotor slots.

3. Press the axle through the rotor until it stops; the rotor should be at the center of the axle.

Section Three - The Frame

The rotor is going to spin on two 608 (skateboard) bearings. You can get 608 bearings at any skateboard shop or on Amazon.com. *Note: Optionally, sandblast the 3D printed frame to take the shine off and make them look a bit more like cast iron.*

1. Press the bearing into both of the frames.

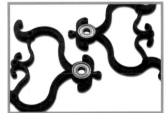

2. Flip one of the frames over, so the bearing is down on the table and press the first electromagnet into place. It's a tight fit, so you may need to turn it back and forth encourage it to go in.

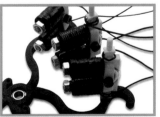

3. Install the other two electromagnets. Make sure to keep the orientations the same.

4. Put the Rotor into place.

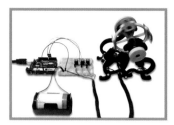

5. Position the other frame and press it into place. The holes for the electromagnets supports are very tight, so you may have to twist the electromagnets back and forth to get the to go into place.

6. Press the timing wheel and pulley onto the ends of the axle.

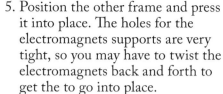

5. The original patent model used a mechanical system to pulse the coils in the right sequence to turn the motor. We are going to replace that part with an Arduino and 3 N-Channel MOSFETS. It will be simpler to build, and easier to tinker with.

6. The electromagnets need 1.5 volts and lots of current, both of which are drawn from a D-battery. The Arduino sequences the MOSFETS, which switch the coils on and off.

Note: *You can get these MOSFETS from Sparkfun:* http://www.sparkfun.com/products/10213

Bread board Diagram

Below is a Fritzing diagram showing how to connect the Arduino, MOSFETS, and motor.

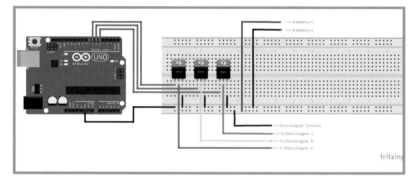

Section Four - Wiring the Magnets

1. Lay out the wires.

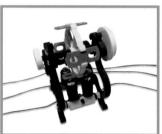

2. Connect the three right-hand wires together to create one common lead. This is the positive lead which connects to the positive side of our controller. The other three leads are the negative leads, which connect each of the electromagnets to the switching MOSFETS. I like to add a section of colored wire to the ends of these three leads to indicate which electromagnet they control.

3. A great way to make the four leads into a single old fashioned feeling power cable is to run them through a shoelace. The wide shoelaces work best. When you cut the ends off, you get a cloth tube.

4. Thread the wires through the cloth tube of the shoelace, and secure the ends of the fabric with heat shrink or electrical tape.

Arduino Code

Below is the Arduino source code for the motor. Load the following sketch onto your Arduino

Note: If you aren't familiar with Arduino, see https://www.arduino.cc/en/Guide/HomePage *for a getting started guide, or check out* <u>Getting Started with Arduino, 3rd Edition</u> *by Massimo Banzi and Michael Shiloh.*

```
/// A simple sequencer for the electromagnetic motor model
int coilA = 7;   // coil array pins
int coilB = 6;
int coilC = 5;
int debug = 13; // Debug LED; pin 13 on most Arduinos
int debugState = LOW;
long previousMillis = 0;
int tarRPM = 10;          // Target RPM for Motor, Changing this value
// with change the motors speed
int pulselength = 0;      // Length of pulse needed to match target RPM,
// value is set in setup.
int coil = 0;             // Stores the next coil to fire
void setup() {
  pinMode(coilA, OUTPUT);
  pinMode(coilB, OUTPUT);
  pinMode(coilC, OUTPUT);
  pinMode(debug, OUTPUT);
  pulselength = (6000 / (tarRPM * 12)); // Sets the pulse length to 6000 millis
  // per min / (target RPM * ((# of Coils)
  // * (# of Cores))
}
void loop() {
  unsigned long currentMillis = millis();
  if (currentMillis - previousMillis > pulselength) {
    previousMillis = currentMillis;
    if (coil == 0)
    {
      coil = 1;
      digitalWrite(coilA, HIGH);
      digitalWrite(coilB, LOW);
      digitalWrite(coilC, LOW);
    }
    else if (coil == 1) {
      coil = 2;
      digitalWrite(coilA, LOW);
      digitalWrite(coilB, HIGH);
      digitalWrite(coilC, LOW);
    } else if (coil == 2) {
      coil = 0;
      digitalWrite(coilA, LOW);
      digitalWrite(coilB, LOW);
      digitalWrite(coilC, HIGH);
    }
    if (debugState == LOW)
      debugState = HIGH;
    else debugState = LOW;
    digitalWrite(debug, debugState);
  }
}
```

Starting the Motor

To get the motor spinning, you'll need to give it a bit of help. The sequencer is very simple, and doesn't have any way to accelerate the motor from a dead standstill. Flick the motor lightly with your fingers to get it spinning. This takes some practice; you need to flick the motor to spin at fairly close to the sequencer's target RPM setting.

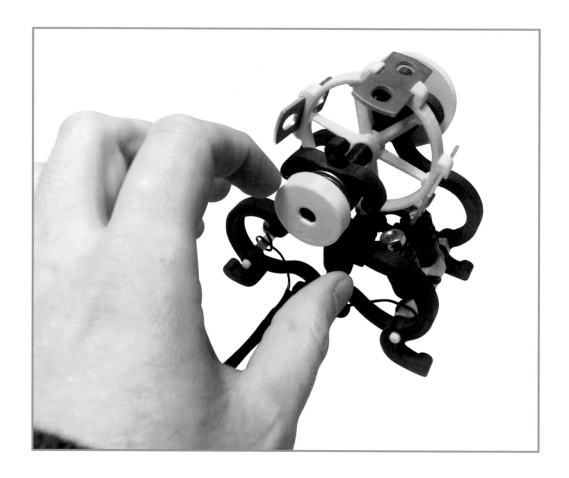

Reversible Lounge

The inspiration for this model comes from Samuel Lloyd's Reversible Lounge (Chapter 8, Home, patent #62,349). To reverse the orientation of the lounge, you simply move its head. No tools are required, and this design allows for equal wear and tear on both sides of the lounge.

HOW TO REPRODUCE THE REVERSIBLE LOUNGE

REPRODUCTION BY KACIE HULTGREN

Bill of Materials

SUPPLIES
- Wood Glue
- 3/16" Plywood, at least 12"x18"
- Screw Eyes or Wood Screws with Washers (2)
- Super Glue
- Felt or Fabric Scraps (optional)
- Cotton or Polyester batting (optional)
- Paint, Shellac, Wood Stain (optional)
- Masking Tape (optional)

TOOLS
- Laser Cutter
- Needle Nose Pliers
- Small Paint brush
- Carpenter's Square or Cutting Mat
- Binder Clips or Hobby Clamps
- Hobby Knife
- 3D Printer (optional)

Part 1: Building the Frame & Headrest

1 Laser cut the pieces for the reversible lounge using the DXF file, available from https://github.com/InventingABetterMousetrap/lounger
The blue lines should be etched, and the red lines should be cut through.

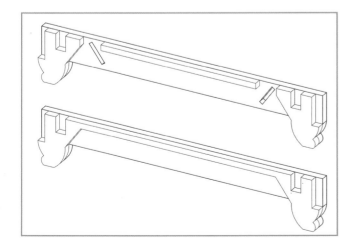

2 Find the pieces that form the back and front of the lower frame. Laminate them together by applying a thin layer of wood glue. Use binder clips or small hobby clamps to press pieces together while drying.

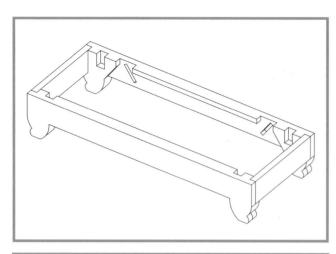

3 Complete the lower frame by gluing together with wood glue. Double check that the frame is square by comparing to the guide lines on a cutting matt or using a carpenter's square.

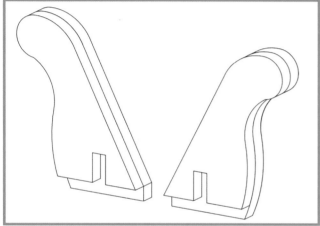

4 Find the pieces for the headrest. Laminate two external pieces, with two interior notched pieces. Use binder clips or small hobby clamps to press pieces together while drying.

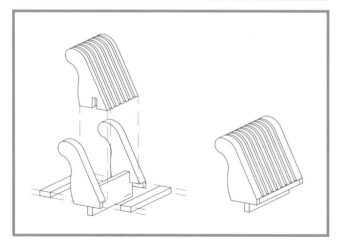

5 Glue the the two cross pieces in place, then slot additional headrest profiles into the cross pieces. Adhere with glue.

Part 2: Wood Finish

If you want to paint or finish the wood pieces in your project, now is the time. Use paint, shellac, wood stain, or a finish of your choice. The lounge in the photo was finished with two coats of clear shellac.

Brush finish on all sides, including pieces which have not yet been assembled.

Part 3: Add Fabric

Adding upholstery for your lounge is optional, and there is more than one way to tackle the project depending on your sewing abilities.

You'll need fabric shapes cut or sewn to the following dimensions: 2 pieces at 8-3/8" x 4" for the seat cushion and headrest, and a piece cut to fit or slightly smaller than the shape of the backrest. Felt is a great choice because the edges don't fray.

Alternatively, create small pillows for each shape. Use 1/4" batting inside, and feel free to sew by hand or use a sewing machine. Use wood glue to adhere the pillows to the appropriate laser cut pieces. Masking tape can be useful to hold the fabric to the wood while it dries.

On the headrest, wrap the excess to the bottom and glue in place.

Part 4: Building the Cushion Frame

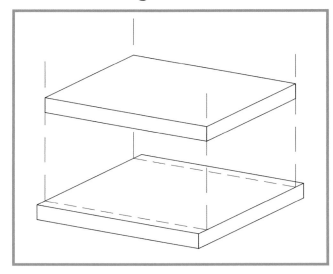

1 Find the three pieces that create the cushion slider frame. Glue the two bottom pieces together, using the etched line as reference.

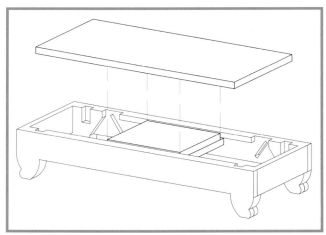

2 When dry, put the lower pieces into the sofa frame as shown.

3 Glue top piece to the slider frame. Hold in place until glue is dry.

Part 5: Install Lounge Back

Locate the lounge back and 2 screw eyes.

Use a hobby knife to pierce the fabric on the lounge back where the guide holes are located.

Twist the screw eyes through the slot in the lower frame and into the holes in the back of the lounge. Use needle nose pliers for additional torque if needed. Screw eyes should finish perpendicular to the slot, and be loose enough that the lounge back can tilt side to side freely.

Part 6: Create Brackets

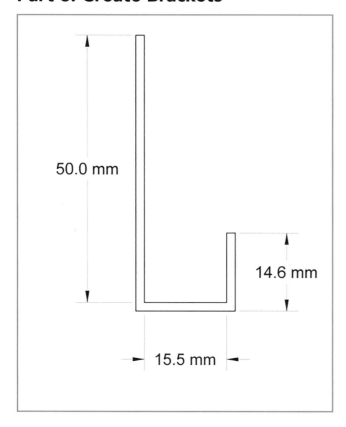

The lounge design requires two brackets to keep the back upright. The bracket for the replicated lounge was created with a 3D printer.

- Print two copies of the *bracket.stl* file, or use the measurements in the diagram below to create brackets of your own with other tools and materials. Glue the brackets to the frame with super glue and let dry. You can download the file from https://github.com/InventingABetterMousetrap/lounger

The Mousetrap

The design for this mousetrap (Chapter 18, Miscellaneous, patent #102,133) featured a unique way for capturing multiple rodents humanely. This particular invention may not have caught on, but its function and construction are a tribute to the seemingly never-ending challenge of building a better mousetrap. Let's first understand how this particular trap works. The contraption is fitted with a rotating table that has four duplicate bait stations, one constructed on each exposed platform. Here's how the mechanism works: first, bait is attached to the top of all four bait hooks which are attached to each bait station on the rotating platform. The trap is then set out to await its first visitor, which cannot resist the tempting treat. The passing rodent, lured by the bait, jumps on the table and pulls on the bait in an attempt to remove it. As the bait is pulled down, the center of the bait hook pivots on a fastener rod and applies pressure to a spring that is positioned on the center of the lower half of the hook. As the bait hook is drawn down by the weight of the rodent, a connected sliding bar begins to slip backward. This rearward movement causes the release of the rotating table to turn due to both gravity and the weight of the creature. As the table rotates, the pest falls into the box, then, as the table continues to rotate around, a duplicate trap will be reset and ready for the next unfortunate visitor. Therein the pest, or better yet, the pests, will reside until they are set free through a bottom door.

HOW TO REPRODUCE THE MOUSETRAP
REPRODUCTION BY DAN CICHELLO

Bill of Materials

SUPPLIES
- Wood selection sizes 1/2 and 3/8 to accommodate the overall construction (most private specialty lumber yards sell larger width quality wood types and can provide professional dimensional planing!)
- Small set of hinges
- Handle
- Locking door hook
- At least 25 1" brass/steel wood screws
- At least 30 1/4" wood screws
- 1/4" wide by 20-22 gauge brass strips (sliding catch)
- 1" 20-22 gauge brass strips (sliding catch covers)
- 1/16"x12" brass rods (for springs) at least 8 (this includes some extras just in case)
- .081x12" brass rods (bait hooks) at least 8(this includes some extras just in case)
- 1/8"x12" brass rod (fasteners for bait hooks) at least 3 rods
- Arbor 1/8 steel rod (at least 6") or use an old 1/8 drill bit
- Box of 1/4" wood tack staples (type to be tapped with hammer)

TOOLS
- Dust mask, safety glasses, and gloves
- Small adjustable protractor
- Cordless drill and drill set
- Small hammer
- Heavy duty wood epoxy or two part epoxy
- Wood glue
- Selection of wood clamps
- Scribe for metal and wood
- Sharp pencils
- Soldering iron and solder (use plumbers solder and flux, not rosin core electrical solder)
- A powered rotary tool (such as a Dremel)
- Various wood and metal bits for rotary tool
- Quality tape measure
- 6" and 12" ruler
- Pair of carpenter squares (12" L-shape)
- Metal cutting shears
- 2 sets of small and medium locking pliers
- Drill bit set (TOOLS, continued next page)

TOOLS (CONTINUED)
- Scroll and band saw
- Various wood files and sanding paper
- Light and heavy duty needle nose pliers
- Small stationary belt sander
- Drill press (optional)
- Medium size table sander for larger panels (optional)
- Spring loaded auto punch (optional)

Here are few things to think about before starting this project. The single most important step is to always wear appropriate safety glasses, gloves, and a good dust mask when working with wood dust and power tools. Next, the careful selection of wood that is nice and straight is necessary for a successful build so check each piece and select pieces with the least amount of warp.

As you move along through the project, some dimensions may need to be slightly adjusted to accommodate different types of wood materials you've chosen for the project. The wood used for my first build was walnut, but you can use just about any wood. Use your best judgment and always plan ahead to the next step to avoid any pitfalls.

Accessories
Many of the pieces needed to construct this trap, like the small hinges and hook latch for the bottom door or the transport handle on the top, can be found in most local hardware and craft stores. If you are looking for a challenge, then everything needed to make this trap can be custom created using a bit of imagination and engineering.

Brass Flats and Rod
Most hobby stores supply brass rods and thin flats that are needed for this project, but I highly recommend acquiring the brass flat and round rods from K&S Engineering. They have been around for quite some time and offer affordable high quality small dimensional brass stock as well as other metals used in hobby, craft, and art. Some stores carry the K&S brand, but if you can't find one near you, check the K&S website which is very informative and intuitive.

Making Springs
Some custom made springs are required in order for this mechanism to work. The springs aren't difficult to make and they don't necessarily have to be as perfect or pretty as the drawings show,

they just need to do the job. Let's go over what type of spring is needed and how to construct them. First, it's best to refer to the drawings in this chapter to better familiarize yourself with the spring's design.

Each set and release mechanism on the rotating platform will rely on a torsion spring, which is a spring that works by torsion or twisting. These springs should be made using 1/16" brass wire rod. You will need to wind these torsion springs by hand around a small diameter rod which will act as an arbor. In this case, the arbor is simply a 1/8" hard round steel rod that can be clamped in a bench vise so the brass wire can be wrapped around it to form the spring coil. Old dull drill bits can make great arbors and come in many diameters.

To make a torsion spring, first find and mark the center of a 8" long piece of 1/16" brass rod.

Depending on whether you are left or right handed, you'll form the spring at the very edge of one side of the bench vise's jaw. The trick here is that you will need to position an 1/8 arbor rod (in my case, an old dull 1/8 drill bit) horizontally in the vise so that at least a 1" tail end sticks out of the side of the bench vise jaw, while simultaneously placing the center of the 1/16 rod vertically against the arbor in the vise and pinch the two together. This will securely anchor the brass rod so that you can now wrap it around the arbor pin.

Use a small set of locking pliers to better grip the brass wire as you begin to wind the brass rod around the arbor. Remember, the brass rod needs to be at the very edge of the vise jaw so you can wrap the wire around the arbor rod while it is pinched in the vise. Wrap it a tight as possible making at least 4-5 coils, but be mindful that the arbor can slip out of the vise jaw if too much force is applied. Try to make the coils as close as possible.

Warning: Be careful: the spring is under tension!

The spring legs on either end of the formed spring coil body need to be in the proper position for the spring to function properly. Use your protractor to verify the required position. Refer to the drawings in this chapter for a visual representation of the position needed for the particular spring being created. The desired position can be achieved by stopping the wrapping of the coil around the arbor. Try practicing by making a few test springs to get a feel for it using scrap or even thinner wire that you may have lying around.

Once you have created the spring body, release the vise jaws and use a set of pliers to re-shape the end that was pinched in the vise. The brass is pretty flexible, you can use the pliers or vise jaws to re-shape the coil body or change the position of the spring legs. Be sure you leave extra wire on both leg ends of the spring to cut down and shape later on in the project.

There is another component for this device that you will need to construct later called the bait hook. This part is made of a slightly thicker rod and arbor but only needs to have a loop made in the center. To create the bait hook component, use the same arbor wrapping technique to create the center loop of the bait hook. The goal is to create a loop to act as a pivot point, not so much a spring. This will be covered in more detail when it's time to make it.

Creating the Body Panels

Be sure your work bench is flat and clear of any debris as you work. Refer to the drawing for the body panel dimensions and carefully lay out and cut the front, back, and two side panels for the body

of the trap. Use a carpenter square when aligning the box—two squares are even better. Remember to mark each panel with a location and label which side is front and back as you progress.

The two side panels should be exact copies of themselves if placed together and stood upright on a flat surface. If they are not, then this error will only come back to haunt you when aligning the center revolving platform. It helps to have a perfectly plotted template of the 10 x 8-1/2" body dimension on a piece of paper. You can set the box down on top of the template for a quick reference as you prepare to join the four panels. The body must be perfectly square in its dimensions; if the body is skewed in any way the revolving door will most likely bind up on the side panels as it spins.

Once you are satisfied with the cut pieces, and before you begin to fasten them together, refer to *Figure 2* and *Figure 3* to locate the areas and dimensions that need to be routed out to allow passage of the revolving door's sliding catches that will protrude outward. These grooves can best be created by using a hand held power tool such as a Dremel and the tool's appropriate wood cutting bits. To route the channels in the wood, first measure and mark the area to be cut, then gradually rout out the wood using the power cutting tool. Be sure to make the channel as smooth as possible by sanding the channel after it has been cut out.

Figure 2:
Front and side panels.

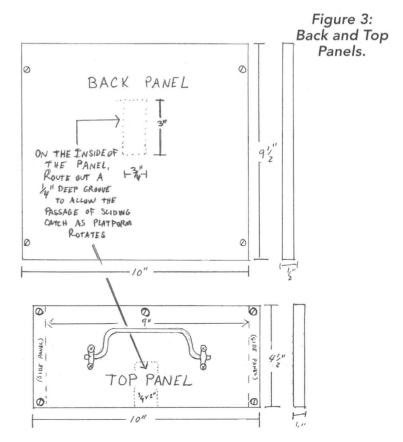

Figure 3:
Back and Top Panels.

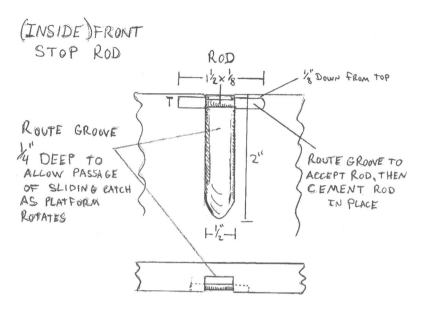

Figure 4: Inside the Front Stop Rod

The front panel will need to be fitted with the stop rod. This rod is there for the sliding catch to rest or stop upon. When the mechanism is activated, the sliding catch will be withdrawn away from the stop rod allowing the rotation to begin. As the table rotates, the next station will have its sliding catch in the set position and will land on the stop rod stopping the table's rotation. Refer to *Figure 4* and route out the section described.

After cutting the slot, cut and insert a 1-1/2" x 1/8" steel rod into the channel and be sure the very edge of the rod is flush with the inside surface area and the top of the rod is 1/8" deep down from the top. Next, secure the rod with an epoxy filler to hold it in this position permanently.

Joining the Body Panels Together

Once you are satisfied with the four body panels, then it's time to join them together. Have a cordless drill with the appropriate size pilot drill selected to work with your 1" wood screws. Attach the front panel to the two side panels and check the squareness before attaching the back panel. Once all four pieces are together, double check the squareness once again before cutting the bottom panel.

The bottom dimensions on the drawings may be slightly different than yours but should be quite close. It's best to cut the bottom piece to fit with your box so that it's as square as possible. The box body will secure and conform to this bottom panel when screwed tight so a custom fit may be necessary to keep everything square. You need to set the bottom base panel at least a 1/4" up inside the box to accommodate the protruding hinges and door lock from scraping the surface that the trap will be set upon.

Bottom Door

Once the bottom is cut and test-fit, remove it and refer to the dimensions in the drawings and cut out the hole for the door by using a scroll saw or by carefully using a jig saw. Use a wood file to get the corners nice and square before cutting out the wood piece to act as the swinging door.

Cut out the actual piece to be used as the door and be sure it has a small amount of play on all sides when set into the bottom panel's door hole so that it can be opened easily. Add the appropriate hinges and hook lock to complete the door then reinstall the base panel and locate the areas that need to be drilled out for the attachment screws and secure it.

Top Panel

Cut the top panel and be sure it is square with the body and lays nice and flat on the top perimeter. Refer to *Figure 3* to locate the area on the inside top panel that needs to be routed out. This channel will allow the passage of the rotating panel's sliding catch. Also, locate the areas that need to be drilled out for the attachment screws. Attach the top panel and be sure everything is square. The top handle could be added now or later if you wish.

Rotating Platform (AKA X-shape table)

Now is a good time to verify and measure the opening on the front of the box were the rotating platform will be eventually be set in. Refer to *Figure 2* and check to confirm your box's 3-1/2" X 3-1/2" measurements, as well as the 9" wide dimension are accurate at the front, top, and mid-point corners before building the rotating platform.

The rotating platform should be made from the straightest wood panel you have and purchased at the 3/8" dimension or planed down to it. Many local wood shops can do this for you fairly cheaply if you have trouble locating this size off the shelf. If the wood you choose to use is too thick, there is a chance the weight will be to great for the rotating action to take place.

Refer to *Figure 5* and accurately draft in your cut lines on the wood panel.

Once the panels are cut, remove the top panel and set each one inside the box to be sure there is at least a 1/16" gap on both sides of the panel.

The rotating table will be in the shape of an X when completed. To bring these two panels together, cut a slot on each panel that allows the panels to slide into themselves. When cutting the joining slot, staying on center is crucial.

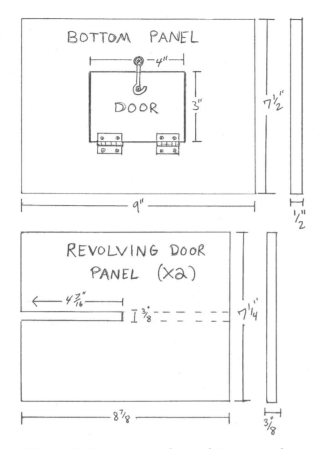

Figure 5: Bottom and revolving panel

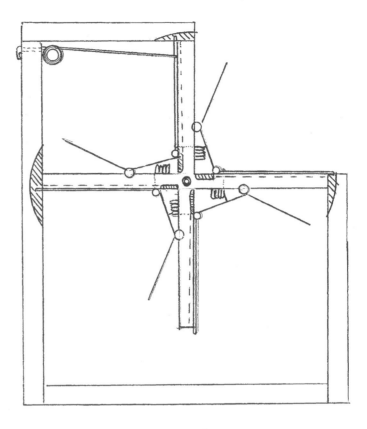

Figure 6: Side View

Remove the top panel and set the X pattern platform into the box and verify it fits by rotating it to all for stations by hand. Verify that you do have at least 1/16" gap on both sides. If you are satisfied with the fit, wood glue the two slip joint panels together and allow to dry.

Now it's time to locate and drill out the center axis of the X shaped rotating table on both sides. On each end of the X rotating table find the center of the X and drill a small pilot hole about 1/2" deep with a 1/16" drill.

Calculate the places on the outside of the side panels where the axis screw will be drilled and drill using the same drill size. Refer to *Figure 6* for a visual.

To test the accuracy of your holes that will serve as alignment of the axis, use two 1/16" x 2" brass rods to act as testing rods/pins that will slip snugly into both pilot holes while positioning the rotating table into the box. If your measurements were correct, the table should rotate freely in the box and spin on your test pins. Since the holes were no more than 1/16", there is a bit of room to make any corrections before you increase the size of the holes to accommodate the actual axis screw to be used as the axis pins.

Center Axis Screws

Once you are satisfied with the rotation, then it's time to create the center axis screws or screw pins. Refer to *Figure 7* to create these.

The drawings will show that the very tips of the screws have had the cutting threads sanded/ground off leaving a smooth tapered point. These points will slip into the pilot holes of the rotating table and the remaining threads will cut into the hole made on the outside panels and will self-tap as they were designed. If need be, you may increase the hole size in the rotating table a bit to accommodate your screw pin's tips, but do this carefully.

The holes in the outside panels will now need to be increased to the appropriate sized pilot holes for your screw's threads to easily anchor into the panel. It's very important to punch an identifying mark on the left and right screws as well as the X shaped table's left and right sides!

Run the screw through the hole to tap out the threads making sure you do this as strait as possible. Remove the screws and set the rotating table into the provided space and visually lining it up to the new holes.

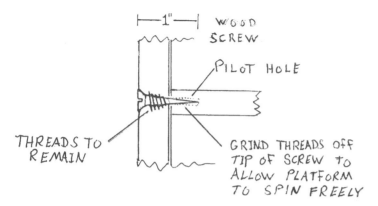

CENTER AXIS SCREW FOR REVOLVING PLATFORM

Figure 7: Bottom and revolving panel

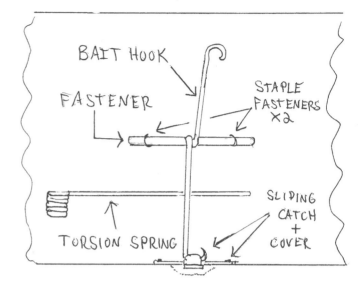

FACING FRONT

Figure 8: The mechanism, facing front

Insert the screws carefully and in a reciprocating fashion, tighten one side a small amount, then the other until they are snug. Do not over tighten, these screws only act as an axis point for rotation. Reattach the top lid and test spin the rotating table and if any of the panels rub or bind while rotating then mark the areas in pencil, remove the table and sand down the high spots until it spins freely.

Catch and Release Mechanism

Now it's time to make the various parts that will act as the catch and release mechanism. But let's examine the contraption and let us review it once again in action while also referring to the relevant drawings to give you a good understanding of the individual component names and their functions.

The mechanism, *Figure 8*, works like this:

• Bait is attached to the top of the bait hook.

• Once the creature pulls down on the bait in an attempt to remove it, the center of the bait hook pivots on the fastener rod and applies pressure to the torsion spring which is positioned on the center of the lower half of the bait hook.

• As the bait hook is drawn down by the weight of the creature, the sliding catch, which when at rest is positioned upon a stop rod located at the front, begins to slide inward. This action causes the release of the rotating table to drop and turn due to gravity, the weight of the creature, and the sliding catch which is no longer resting on the stop rod due to its retraction.

• As the table rotates, the pest falls inside the box and the torsion spring again pushes the sliding catch forward so that it can later land on the stop rod.

• As the rotating table continues to rotate around it presents a duplicate trap reset and ready for the next unfortunate visitor.

It's best to remember there are four separate platforms and every component may have a degree of variation so it's very important to keep each set separated and organized to go with its own platform.

The first piece to create will be the sliding catch cover plate. The sliding catch plate should be made out of 1" flat stock (approximately 20-22 gauge thickness) and cut down to 2-3/4" long. Refer to *Figure 9* to locate and drill out the holes needed for the entry of the small wood screws that will eventually be needed to fasten the covers down to the platform.

The sliding catch is made from a strip of 1/4" wide by 20-22 gauge metal. Cut it longer than the dimension on the drawing to allow you room to bend over one end to form a loop and additional length on the other end which will be cut to proper length later on.

Using heavy duty needle nose pliers, carefully form a loop on one end of the strip.

Use a good soldering iron to heat and melt a small amount of solder to close the gap were the bent loop meets the flat to help prevent the loop from opening. Again, wait until all the other pieces of the mechanism are made before cutting the newly constructed sliding catches down to proper length.

In order for the bait hook to function properly and prevent it from wobbling as it moves, it needs to be mated with a fastener that the bait hook can freely pivot around, and at the same time hold it securely to the rotating table. The fastener will need to be 1/8" brass rod and the bait hook's pivot loop will need to slip over this rod.

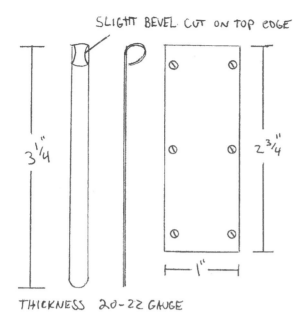

Figure 9: Sliding catch and cover

To form the loop in the center of the bait hook blank you will use the same technique as the one used to form springs in the bench vise; but this time use an ⅛" steel rod arbor or drill bit. You can use the 1/8" brass rod as an arbor but it may distort due to the brass being a much softer metal.

Warning: Be careful; the metal is under tension during this step.

Refer to *Figure 10* to visually see the 135 degree position required when coiling the .081 rod around the 1/8" arbor, then cut a straight blank at least 6" long and position the arbor in your vise.

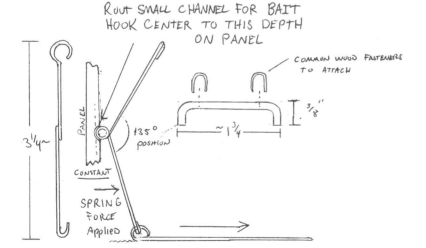

Figure 10: Bait Hook and Fastener

Hand-wind one revolution around the arbor to a 135 degree position. Use a protractor to check the position and fine tune it by hand if necessary.

To bend the bottom hook, which will connect to the sliding catch's loop, cut the blank down to 2" long on one end. This hook needs to slip into the loop of the sliding catch piece already constructed so the hook width must be large enough to accommodate it. Form the hook using heavy duty needle nose pliers.

The top portion of the bait hook will need a larger loop hook for attaching the bait. You can be creative here, just be sure the length from the center pivot to the top of the bait hook is no longer than 1-3/4 overall.

To make the fastener staple which will both hold the bait hook to the panel and also allow it to pivot, first refer to the drawings for length and use the 1/8" diameter brass rod.

To anchor the fastener into the wood panel you will need to form legs, similar to a staple's legs. The inside span of the staple should be close to 1-1/2", so cut a piece of 1/8" brass rod to no less than 3" long, mark the rod at the bend points, and gently peen the legs over in a vise to 90 degrees.

Warning: Metal gets hot very quickly when sanding or grinding on a machine so use gloves and go slow.

After bending the legs, cut or grind down the legs to 3/8" overall. These cut legs will eventually be anchored into the wood and then the fastener will be attached using two common heavy duty wood fastener staples.

The torsion springs, discussed earlier, will now be needed. Be sure to always leave plenty of wire on both ends of the spring to cut down to appropriate length as you fit and assemble. The leg position of 100 degrees is most important! Use a protractor to check the leg position and refer to *Figure 11*. You want the spring to always apply outward force against the bait hook below its center to keep the sliding catch pushed forward.

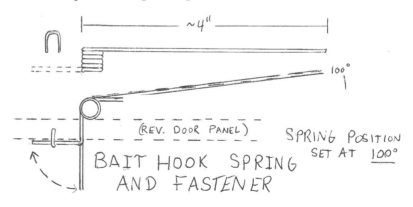

Figure 11: Bait Hook and Fastener

Once all the springs, bait hooks, and other components are made it's time to start mounting them onto the rotating platform. Remove the X pattern rotating table out of the box and choose a platform to start adding components to.

Set a cover plate over the sliding catch's groove making sure the edge is positioned at the very edge of the wood platform. Drill small holes for the wood screws to attach the cover.

Remove the cover and place one of the sliding catch components in the groove. Using small 1/4" long wood screws, reattach and screw down the plate to cover the sliding catch thereby trapping it in the groove.

Visually spot the best position for the center of the bait hook to pivot. Mark this area with a pencil and route out a small channel just wide enough for the center loop to set into. As the pieces are assembled, always verify that the slipping action of the sliding catch is maintained and be sure it will not bind to much when set underneath the cover plate. Refer to *Figure 12* and note the depth of the channel to be cut.

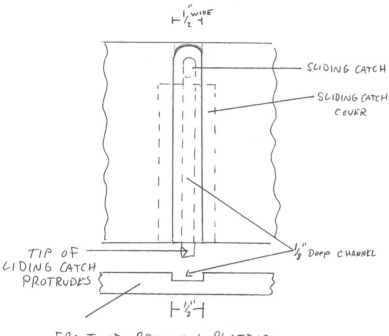

FRONT OF REVOLVING PLATFORM

Figure 12: Looking Down at the Revolving Platform

After cutting the small channel, slip the fastener staple through the center of the bait hook's central loop. Position the staple and mark the best location to drill out pilot holes for the fasteners legs to be lightly hammer-tapped into, which would then hold the bait hook against the wood and allow the bait hook to pivot freely. It's important to lay the panel that's being hammered flat against a

stable surface to help absorb the blow and minimize any distortion to the X pattern's shape as it's being worked upon.

Now check to be sure that when the bait hook is moved up and down that the sliding catch moves freely in its channel in response. Once it passes this test, then you can use a small amount of two-part epoxy to assist in anchoring the fastener into the wood if you feel it is too loose. Tap in two small wood tack staples to firmly attach the fastener to the wood.

Next, set the rotating table back into the box with the pivot screws and observe that the sliding catch's length will need to be cut down. Push upwards on the bait hook to push the sliding catch to its maximum outward length. Cut the excess length off the end so that enough overlap will be present to come into contact with the stop rod previously installed, while also making sure that when the bait hook is pushed downward that the sliding catch can retract enough to allow it to disengage with the rod allowing rotation. Repeat this process on the remaining three platforms.

The final piece to add to each station on the platform is the torsion spring which is required to push and hold the bait hooks outward. Refer to the drawings to visually assist in installing the springs.

Remove the rotating table once again and choose a location approximately two inches in from the edge of the panel and you want the spring leg that applies pressure to the bait hook to rest at the center point of the bait hook leg.

Refer to *Figure 10* and *Figure 11* and mark and drill a small hole to snugly slip the back end of the torsion spring leg through until the coil body of the spring rests against the panel. On the back side of the panel, while resting the panel on a hard flat surface, apply pressure on the coil body against the panel underneath and be sure to keep the leg positioned on the center of the bait hook's leg. Now you peen over the back side spring leg and secure it with a wood tack staple (*Figure 13*). Test the torsion of the spring against the bait hook to be sure it is adequate to keep the sliding catch pushed outward and can retract easily when pressure is applied to the spring through pushing down the bait hook. Do the same on the remaining three platform stations.

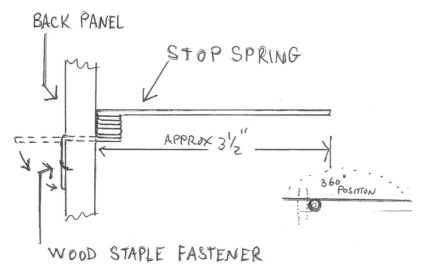

BACK PANEL

STOP SPRING

APPROX 3½"

360° POSITION

WOOD STAPLE FASTENER

Figure 13: Mounting the second stop spring

Carefully peen over the leg and use a wood tack staple to hold its position. Adjust the spring to allow slight contact and cut the spring length so it will be able to drop down behind the panels just enough to prevent backward movement.

If everything is working well, then it's time to reattach the top lid and paint, stain, or wax the finish to complete. If you choose to create your working model of this novel contraption, don't be afraid to implement your own ideas. This replication described here is a duplicate of the very model originally submitted for patent in 1870. It's all about imagination and invention, have fun, and if you can build a better mouse trap, then that's what needs be done! Good luck!

Stop Spring

To prevent the rotating platform from moving in reverse, you will need to mount a stop spring inside the box. This spring will be similar to the torsion springs created earlier only the position needs to be 0 or 360 degrees, or in other words, both legs leveled flat in opposite directions. Use the same 1/16 brass rod for the spring and 1/8" arbor to coil and form the body around. Four to five coils should be fine.

The spring will be positioned on the inside right or left corner, approximately one inch in from the inside edge and 1/2" down from the top. The idea is to allow the panel to swing around and just skim the spring enough to easily allow it to pass without interfering with it's rotating momentum. At this stage the rotation halts due to the sliding catch hitting the stop rod out in front. Now that the new spring will be installed, it will spring down just enough for the tip of the leg to prevent the rotating table from going into reverse.

Form your torsion spring with extra length on the legs, at least 3 1/2" from the center coil for now. With the rotating table installed in the box , visually check that the leg of the spring coils off the top of the spring's body. Do not set the spring down to far inside when choosing the location point otherwise it will stop the rotation of the table.

Drill a small hole for the back end leg to slip through and cut down to a shorter length protruding approximately one inch from the outside panel.

Pigeon Starter

Patented in 1875, the Pigeon Starter (Chapter 22, Sports & Entertainment, patent # 159,846) was a device used to startle uncaged pigeons for target practice. Over time, the sport, trapshooting, evolved into using inanimate clay disks instead. The original model used spring steel to actuate the model cat. However, because of the scarcity of spring steel and its potential danger if installed incorrectly, we have replicated the mechanism using a much safer option, surgical tubing.

HOW TO REPRODUCE THE PIGEON STARTER
REPRODUCTION BY NATHANIEL TAYLOR

Note: The artist built the pigeon starter with an improvisational style utilizing more advanced and dangerous techniques (such as welding). He recommends following this build as a rough guide and not a recipe. Approach with an inquisitive mind and experiment!

Bill of Materials

SUPPLIES
- Gorilla glue
- 2-part epoxy
- Rubber slingshot tubing (1/4" inner dia., 3/8" outer dia.)
- Heavy duty hinge
- 3/4" Baltic birch plywood (15 ply, void free)
- 4" Balsa wood
- 1/4" linear shafts, stainless steel x 6
- 1/4" shaft collars x 6
- T bar steel channel
- Steel cable
- Metal spring 1/4" outer diameter, compression rate ~ 5 lbs/in
- Assorted screws

OPTIONAL
- Broom bristles
- Doll eyes
- Polyurathane finish
- Walnut tile planks (3/4"W x 1/8"H)
- Various acoustics instruments (toy rattles, cymbals, nuts, etc)
- 1/4" strike plate

TOOLS
- Table saw
- Jig saw
- Metal saw
- Angle grinder
- Hex-keys set
- Wood/metal drills
- Sander
- Metal file
- Dremel with carving bits
- Paint brush (2"-3")

OPTIONAL
- MIG welder
- Metal mill
- Plastic extruder

The Base of the Pigeon Starter

Purpose: The base keeps the legs and hinge mechanism fixed in place. It should be heavy and durable.

1. Cut the 3/4" Baltic birch plywood to 13.5" x 21.5"

2. Cut the Baltic birch plywood for the four legs—about 7.5" x 3" for each. Do not sand the legs down yet.

3. Plan the base layout—center the legs on the board; give about 12" of distance between the leg pivots (aka the cat feet)

4. Make sure the front and hind legs are exactly aligned. This will eliminate unwanted friction and torque later on.

5. Center the hinge between the T-bar rails.

6. Mill the T-bar steel channels. Leave about an inch of steel channel for attaching the wooden legs.

7. Drill holes to fit two 1/4" shafts through the T-bar and legs. (one shaft connects the front legs, the other connects the hind legs). Fix legs in place with 1/4" shaft collars. Drill holes into the base of the T-bar. Screw the T-bar into place.

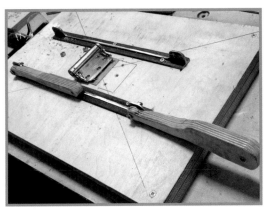

8. Your cat feet should now rotate on the T-bar rail, which is fixed to the base.

The Body of the Pigeon Starter

Purpose: The body should be as light and hollow as possible. This allows space for instruments while keeping the load light.

BOX FRAME OF CAT BODY

1. Build a sturdy box frame with dimensions that will fit between the legs. (Optional) To give the cat a more organic shape, create grooves in the box frame where it attaches to the legs.

2. Like the leg pivots, use two 1/4" shafts to connect the front and hind legs to the box frame.

3. Secure in place with 1/4" shaft collars.

MAIN BODY

1. Outline the body of the cat with balsa wood. To increase the body height, glue two blocks of balsa together.

2. Shape the cat with a Dremel tool, sander, etc.

3. Outline the cat cavity and hollow it out as much as possible.

4. Coat both sides with epoxy to strengthen the balsa wood and give it a nice finish.

5. Screw the box frame to the main body for easy removal/attachment

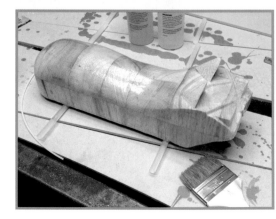

Hinge

Purpose: The hinge is the most important component of the pigeon starter. It generates the force of the "spring" mechanism and also absorbs most of its impact.

This is the finished version of the hinge, with tiling added to hide the strike plate and T-bar railing.

1. Screw one part of the hinge in place onto the base.

2. Cut a 4" tall wood block (from Baltic birch plywood). *The wood block prevents the hinge from traveling past 90 degrees.*

3. Screw the mobile part of the hinge to the wood block.

4. Cut the T-bar channel to fit the length between the hinge and the box frame.

5. Screw the T-bar channel on the wood block, right above the hinge. *Replace screws with nuts and bolts for reinforced*

strength.

ATTACHING THE HINGE TO THE CAT FRAME

1. Plan out where the T-bar will connect to the box frame. Drill holes for the 1/4" shafts, and mill out the rest of the sides of the T-bar. *Because the cat will be moving both horizontally and vertically, drill larger holes to accommodate tolerance for movement.*

2. Drill holes on the T-bar for the rubber tubing. *Drill the tubing holes as close to the box frame as possible without impeding it.*

3. Connect the T-bar to the sides of the box frames with a 1/4" shaft.

CREATING THE "SPRING" TENSION FOR STARTING PIGEONS

1. Drill two holes near the cat's front legs.

2. Thread the rubber tube through the holes so that the two ends evenly touch the

base. *Optional: Slingshot rubber tubing connects a pair of identical lengths with helpful plastic stoppers, which prevent the tubing from moving around. This makes the next step a lot easier.*

3. The rubber tube should be cut to length so that it barely reaches the drilled holes in its relaxed state.

4. To create tension, thread a rubber tube through the hole in the base and tie a

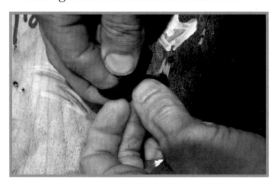

knot.

5. Repeat for the other rubber tube.

6. Adjust tension by pulling the rubber tubes by slackening or tightening the ends. *Note: If you do not have slingshot rubber tubing, pull both tube ends through the holes and tie them together.*

Latch Mechanism

Purpose: The latch keeps the rubber tubing tense until release.

1. Mill the T-bar channel into a shape that supports latching onto the 1/4" shaft from the cat frame.

2. Drill a hole for the release cable.

3. Drill a hole to fit the latch onto a 1/4" shaft.

4. Add a spring behind the latch to keep the latch pushed forward

Optional: To reduce unwanted torque, weld a 1/4" steel tube to the T-bar to keep it stable.

Instruments

The instruments are mostly collected from toys and blended together to make a scratchy, hissing noise. Experiment with different acoustics around you and see what works.

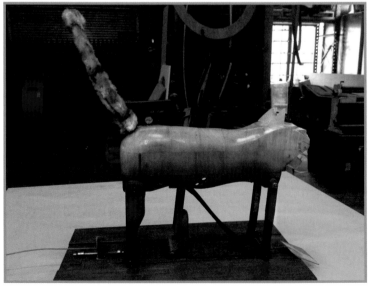

The completed Pigeon Starter model

Rowboat

The original rowboat (Chapter 7, Toys, patent #133,250) uses a toy key to wind up a spring steel mechanism, which then moves gears to actuate the actual rower. In the current version, we use LEGO Mindstorms to power the rowboat with a similar set of gears. As the structure of the boat will vary depending on what LEGO pieces you have on hand, these instructions focus only on the gear mechanism.

HOW TO REPRODUCE THE TOY ROWBOAT
REPRODUCTION BY JAMES FLOYD KELLY

Bill of Materials

You'll need the parts shown in Parts list and the LEGO Mindstorms toy rower program from https://github.com/InventingABetterMousetrap/LEGO-Boat.

PARTS LIST

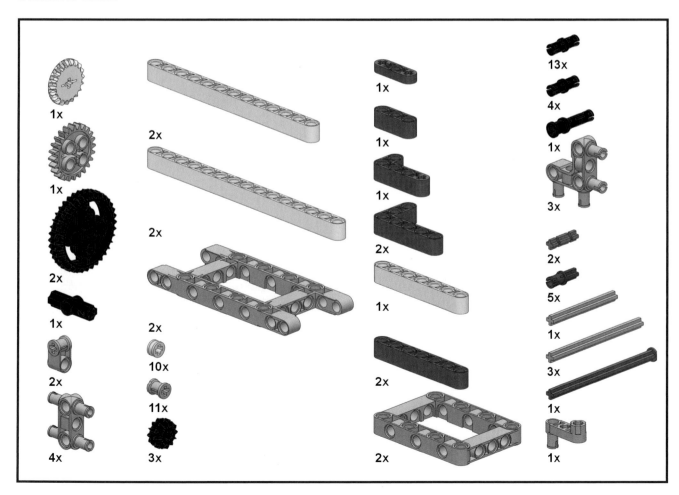

The Gear Mechanism

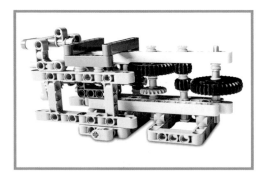

Gear Base

1. Connect four 64179s (Technic Beam 7 x 5 with Open Center 5 x 3) to four 48989s (Technic Cross Block 1 x 3 (Pin/Pin/Pin) with 4 Pins) as shown:

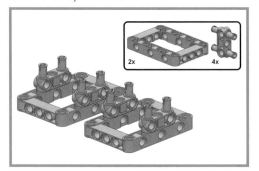

2. Connect four 2780s (Technic Pin with Friction and Slots) and two 41239s (Technic Beam 13):

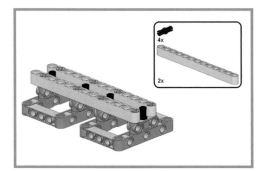

Gear Assembly

3. Combine one 32278 (Technic Beam 15) with one 32123b (Technic Bush 1/2 Smooth with Axle Hole Semi-Reduced), three 44294s (Technic Axle 7), and five 3713s (Technic Bush with Two Flanges):

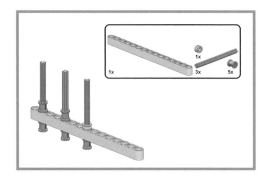

4. Let's put the gears together. Combine one 32498 (Technic Gear 36 Tooth Double Bevel), one 32198 (Technic Gear 20 Tooth Bevel), one 32270 (Technic Gear 12 Tooth Double Bevel), three 32123bs (Technic Bush 1/2 Smooth with Axle Hole Semi-Reduced), and two 3713s (Technic Bush with Two Flanges):

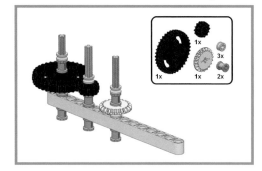

5. Next, combine the Gear Base and Gear Assembly as shown in the next two diagrams:

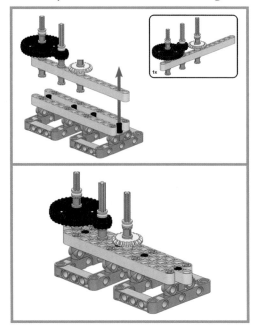

Crank Assembly

1. Begin by combining one 32270 (Technic Gear 12 Tooth Double Bevel), one 55013 (Technic Axle 8 with Stop), three 3713s (Technic Bush with Two Flanges), and one 6536 (Technic Cross Block 1 x 2 Axle/Pin):

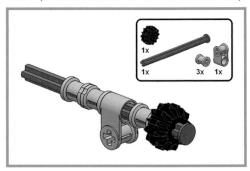

2. Then, combine one 32062 (Technic Axle 2 Notched), one 32034 (Technic Angle Connector #2, 180 degree), and one 55615 (Technic Pin Connector Perpendicular 3 x 3 Bent 90 with 4 Pins) as shown:

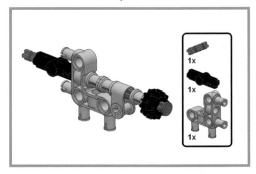

3. Now the crank is starting to take shape. Add in one 3648b (Technic Gear 24 Tooth with Single Axle Hole), one 43093 (Technic Axle Pin with Friction), and one 33299 (Technic Beam 3 x 0.5 Liftarm with Boss and Pin):

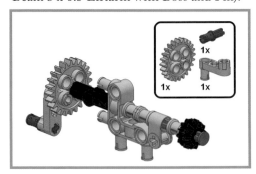

4. Next, add two 2780s (Technic Pin with Friction and Slots) and one 32062 (Technic Axle 2 Notched):

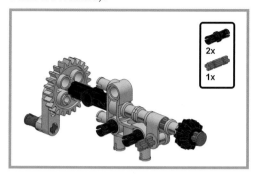

5. Connect a 32140 (Technic Beam 2 x 4 Liftarm Bent 90):

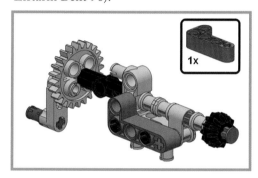

6. Now, attach the Crank Assembly to the Gear Base/Gear Assembly as shown in the next two diagrams:

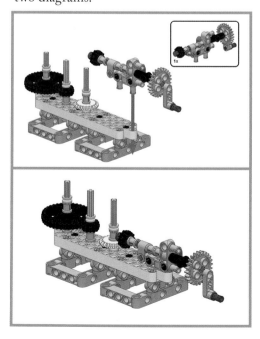

7. Add in one 32270 (Technic Gear 12 Tooth Double Bevel) and 32498 (Technic Gear 36 Tooth Double Bevel):

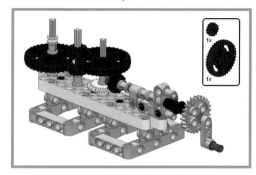

8. Cap it off with two 32123bs (Technic Bush 1/2 Smooth with Axle Hole Semi-Reduced) and a 32278 (Technic Beam 15):

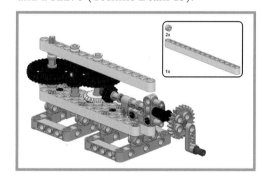

9. Insert four 2780s (Technic Pin with Friction and Slots) as shown in these two diagrams:

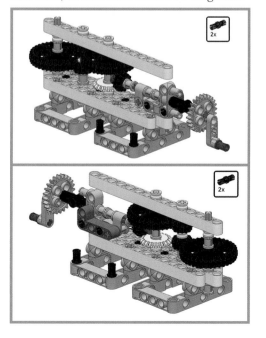

Complete the Chassis

1. Attach a 64178 (Technic Beam 11 x 5 with Open Center 5 x 3) as shown:

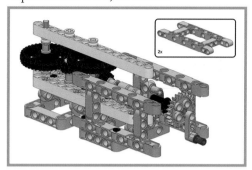

2. Next, connect four 43093s (Technic Axle Pin with Friction) and four 32123bs (Technic Bush 1/2 Smooth with Axle Hole Semi-Reduced):

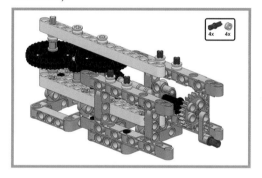

3. Connect two 32526s (Technic Beam 3 x 5 Bent 90) and six 2780s (Technic Pin with Friction and Slots) to the assembly:

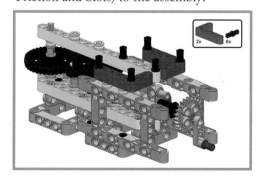

4. Now you'll need to attach two 55615s (Technic Pin Connector Perpendicular 3 x 3 Bent 90 with 4 Pins) and two 32524s (Technic Beam 7) like this:

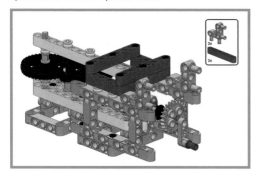

5. Now build up one 3713 (Technic Bush with Two Flanges), one 32073 (Technic Axle 5), and one 6536 (Technic Cross Block 1 x 2, Axle/Pin) into this assembly:

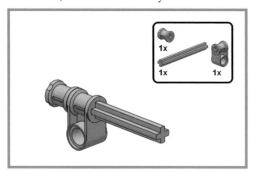

6. Add in one 32523 (Technic Beam 3) and a 32054 (Technic Pin Long with Stop Bush):

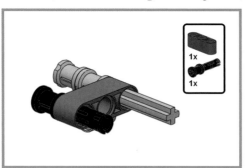

7. Then combine that with one 32524 (Technic Beam 7), a 2780 (Technic Pin with Friction and Slots), and a 6632 (Technic Beam 3 x 0.5 Liftarm):

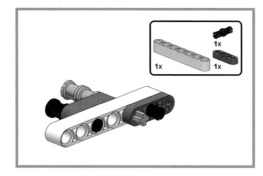

8. Attach this new assembly to the main structure as the next two diagrams show:

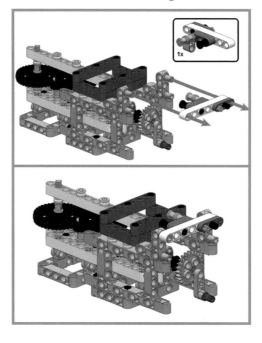

Powering and Using the Gears

1. The motor attaches to the first gear from underneath:

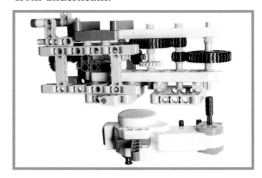

2. The blue pin connector attaches to the oar, which then moves the entire rower:

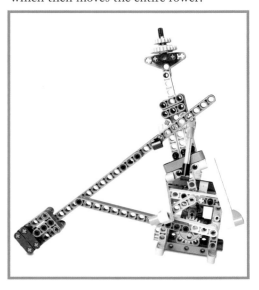

3. The rest of the boat is built around the gear piece as shown on the next page.

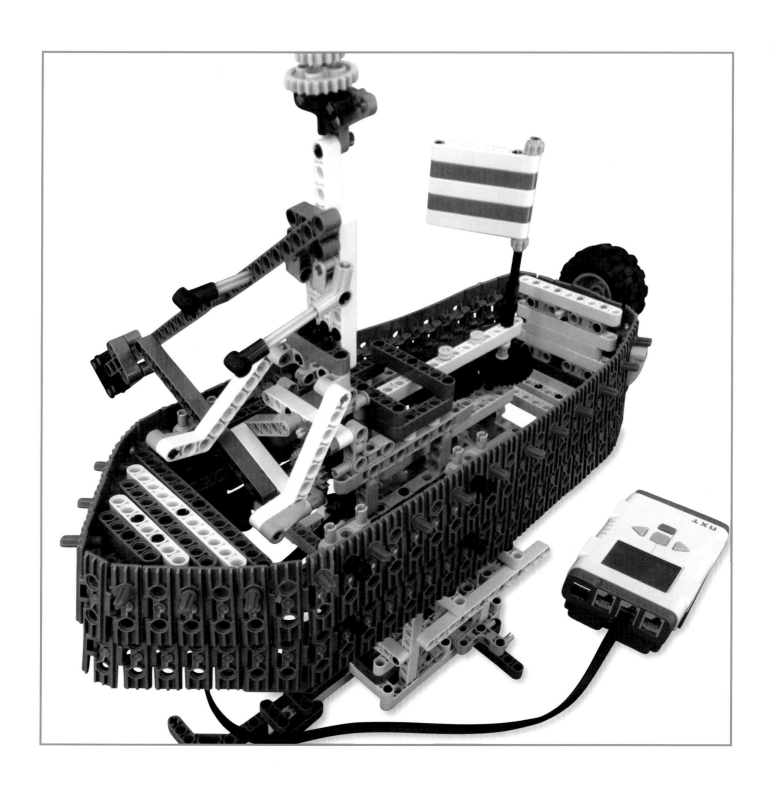

Washing Machine

The original patent model for the washing machine (Chapter 10, Laundry, patent # 90,416) was designed in 1860 as the first washing machine with a water heating mechanism, as hot water was previously carried to the tub. The design of the tumbler is similar to some modern machines. The model has a rotating tumbler made out of interlocking shingles and a hand crank. The shingles trap and release water as the tumbler turns. As the user turns the crank, the coal tray beneath the wagon heats the reservoir of water.

HOW TO REPRODUCE THE WASHING MACHINE
REPRODUCTION BY JUDY AIME' CASTRO

SUPPLIES
- Wagon: 5" x 7⅛"
- 1 piece of wood: 24" x 18" x ¼"
- 1 piece of brass: 5¼" x 14½"
- 2 Wood screws top row: size #4 - 40 x ⅜"
- 2 Wood screws bottom row: size #4- 40 x ⅜"
- 1 box of 100 Escutcheon pins: size ½" x 18 Ga
- 2 brass nuts and bolts #4
- The Tumbler: Overall diameter: 7" x ¼" thick
- Copper sheet: 016 copper sheet of 12" x 18"
- 20 Individual size per shingle: 3¾" x 1¾"
- 35 pieces of bass wood square dowels: 3¼" x ¼"
- 2 door latch: 2" x ⅜"
- 1 Pipe, copper tubing: 8" x ½" diameter
- 1 Pipe, copper tubing: 4" x ½" diameter
- 2 Copper elbow joint: wide angle x ½" diameter
- 1 Skewer: 2"
- 2 Conduit strap: custom to fit
- Hand crank: 4", with 1.5" at the right angle handle
- Door hinge: ¾"
- Coal tray length: 6" x 4"

TOOLS
- Awl
- Band saw/laser cutter
- Fine grit sandpaper
- Hammer
- Hand tools
- Metal cutter
- Metal file
- Needle nose pliers
- Phillips head screwdriver
- Small vice
- Wood glue

The Wagon

Attach the legs to each side of the tub with the brass wood screws. The front legs are shorter than the back. Use a small amount of glue between the screw holes.

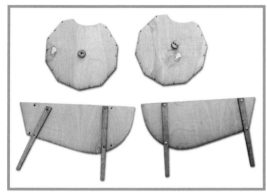

The top row of wagon screws are #4 ⅜" and the bottom row are #4 ⅝" wood screws.

Square all four sides of the wagon and glue all corners. Turn the wagon upside down and use water bottles to prop up the wagon walls.

The Tumbler

The tumbler consists of a door, pairs of interlocking shingles, and a hand crank (for agitating the clothes).

PART ONE:

The inner layer of the tumbler is constructed from supporting square dowels. Pre-make the holes all around the tumbler and use a bit of glue and escutcheon pins to nail both sides of the tumbler to the dowels.

Attach only the corner trusses, don't attach the middle trusses until you are done installing a pair of shingles.

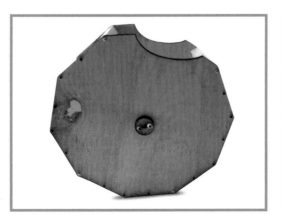

PART TWO:

The outer layer of the drum consists of ten pairs of interlocking copper shingles. Draw all your pieces onto the copper sheet, cut them out, then hammer and file down the sharp edges.

Once you have paired ten sets of shingles, begin marking where the nails will go. Drill small pilot holes on the edges of the copper with an awl first, and then use nails to secure the copper shingles to each square bass wood dowel.

Attach the first set of shingles aligned against the square dowels. Add more dowels as needed.

To keep the shingles flush against the tumbler, add a nail between the shingles and the middle truss. Avoid over-hammering the tumbler.

When you are done attaching all the shingles, attach the wood washer on each side of the tumbler, leaving the middle hole free.

THE BOTTOM OF THE TUMBLER

Cut the shape out of brass using the pattern, then file all the edges and round the corners. Fold the flanges that will hold the tray below the wagon. Use a metal file to bend the flanges down and towards the center. Shape the brass into the roundish shape to fit the tub. Brass bends easily, so it can be shaped over your thigh. Test the shape against the wagon and continue bending the brass until you acquire the desired shape.

Take one pair of shingles and fit them against each facet of the decagon. Additional trimming may be necessary as you fit in the shingles. Always smooth the edges after every trimming. Continue shaping the shingles until they fit flush against the tumbler.

When the first pair fits properly, proceed to the next pair. Always work them as a set. The orientation of the shingles is important as they all must face the same direction.

As these nails are small, use a pair of pliers to squeeze the nails into the bass dowels.

Once you have installed the first set of shingles, install the middle truss, and then proceed to the next set of shingles.

When you have attached almost half of the shingles, the tumbler will be difficult to hold. Rest the tumbler between your knees and lightly press or hammer in the nails.

THE COPPER COAL TRAY

The tray consists of a box and two outward flanges at each side, which allows the tray to slide in and out from from the bottom of the wagon's inward flanges. The tray connects to the exhaust pipe.

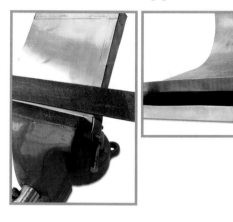

THE EXHAUST PIPE

Construct the exhaust pipe by joining a copper pipe and two elbow joints. Attach the pipe to the wagon by creating a copper conduit strap with couple of nuts and bolts. Use the other strap to support the pipe at the bottom of the wagon.

HAND CRANK AND ROTATING PIN

The tumbler has a hand crank on one side and a rotating pin on the other.

For the rotating pin, cut a piece of skewer about 2" and insert in one side of the tumbler.

For the other hand crank, cut a piece of a wire coat hanger, bend the wire and insert the other end in the middle hole of the tumbler. Use a pair of pliers to make a small hook inside to secure the crank to the tumbler.

THE LATCH

Construct a small metal latch by cutting a piece of brass or copper, shaping and making a hole at each end. Insert a couple of nails on the tumbler, using the head of a nail as the locking mechanism at one end and rotating pin on the door.

THE DOOR

Construct the door with three trusses, two side pieces, and two overlapping shingles.

Finally, attach a small door hinge to the door.

What's Next?

"Invention, it must be humbly admitted, does not consist in creating out of void, but out of chaos."

—Mary Shelley

**BY DAVID COLE, EXECUTIVE DIRECTOR,
HAGLEY MUSEUM AND LIBRARY**

It wasn't a typical Sunday stroll. From 1810 until 1876, residents of Washington D.C. would often incorporate a visit to the U.S. Patent Office into their Sunday afternoon perambulations. Initially at Blodgett's Hotel, and after 1840 at the office's neoclassical headquarters between F and G streets, curious citizens would come face to face with the latest creations of American inventors. In a long exhibition hall lined with imposing glass cases, visitors to the Patent Office gallery were presented with a one-of-a-kind spectacle—an array of thousands of patent models, representing innovations across the commercial spectrum. As an invitation to the public to inspect the latest fruits of American ingenuity, this display foreshadowed, on a small scale, the exhibits of industrial and design achievements at the world's fairs of the late nineteenth and twentieth centuries. For a people whose own nation was, for better and worse, an ongoing experiment, this display would have served as a visual record of an enterprising, dynamic present—and as a harbinger of big things to come.

Those nineteenth-century patent model displays reflected the spirit of a people passionate about innovation and entrepreneurship. Fortunately, that spirit is alive and well in the twenty-first century, thanks in part to a pair of present-day visionaries. Following a century of ill-fated attempts by others to create patent model exhibitions with truly national scope and exposure, Alan and Ann Rothschild have spent two decades assembling and exhibiting the Rothschild Patent Model Collection. The Rothschild Collection, as this handsome and compelling volume attests, is nothing less than a national treasure; its more

than 4,000 patent models, patent certificates, and voluminous research documents chronicle over a century of American history through our inventions. Each of these models has multiple tales to tell—of Americans at work and at play, of innovators grappling with problems in factory yards and in barnyards, and of the ceaseless drive to meet challenges presented by the settlement of a vast and rough country. The stories they embody—economic, social, cultural—are intriguing points of departure for any student of the events and trends, large and small, that shaped the development of the United States.

Just as important, attending to these models reveals how individuals tackled innovation challenges in the nineteenth century and illuminates the experimental methods and problem-solving processes that led inventors from concept to finished product. Examined carefully, one by one and in the aggregate, they also help dispel myths about innovation that might prove discouraging to would-be inventors today. One such myth is that the important inventors of yesteryear all had famous names. To be sure, the Rothschild collection contains its share of models by luminaries such as Corliss, Whitney, Goodyear, Maxim, and Fleischmann. It is also true, however, that the lion's share of the models were produced by people who had been lost to history—and their contributions represent multiple demographics, industries, and geographic origins.

An aspiring innovator could also be excused for believing that inventors are specially gifted people who work alone, and whose world-changing inspirations come to them in "Eureka!" moments. This conceit makes for exciting drama, but the truth (borne out by the Rothschild Collection) is usually less thrilling. An invention is typically the product of long hours (even years) of experimentation by a person whose talents, while impressive, are not extraordinary; it is often a response to a practical problem that addresses a parochial concern in the inventor's "backyard" and does not have global consequences. Moreover, as Mary Shelley observes, the romantic notion of the inventor as a solitary genius who creates "out of void" is at odds with reality. Patentable inventions usually take the form of "improvements" to existing attempts to solve a common problem. The Rothschild Collection demonstrates that inventors, who wish to make a better washing machine, steam engine, or mousetrap, are embedded in a rich environment of inventions and engaged in a kind of conversation across space and time with numerous other innovators. They create, not out of thin air, but "out of chaos."

The Rothschild Collection is an unmatched resource for sharing these invaluable insights into inventors and the culture of innovation. Mindful of its great potential as an educational tool, the Rothschilds and I are delighted to announce that, as of 2015, their Collection has become a permanent addition to the holdings of the Hagley Museum and Library in Wilmington, Delaware. Dedicated to preserving and presenting the history of business, technology, and innovation, Hagley is ideally positioned to tell stories from these fascinating artifacts and to share them in Wilmington and across the globe. Through engaging multimedia exhibitions, digital content, and publications built around these patent models, Hagley will work to raise public awareness of invention and innovation in America, past and present.

In this endeavor, we aim to recreate the spirit of those Sunday afternoon strolls through the Patent Office that thrilled citizens more than 150 years ago. Our hope—and expectation—is that new generations of inventors and entrepreneurs, inspired by the ingenuity and creations of their forebears, will be encouraged to ask themselves the question "What's next?" We look forward to the fruits of their inspiration.

Index of Inventors

This index lists all the inventors whose models are described in this book.

Index of Models